Computational Fluid Dynamics: A Powerful Tool for Understanding Complex Flows

Nami

Contents

1 Introduction

In the past decades, numerical simulation has become an essential tool in research and engineering applications. The further development and availability of computational resources has enabled the use of advanced software solutions as a well-grounded alternative to experiments and theoretical analysis. In this context, Computational Fluid Dynamics (CFD) methods have gained great popularity in the investigation of complex fluid flows, providing a way to generate output data at points of interest, where experimental measurements are impossible or theoretical analysis is limited, as well as faster available results and lower costs.

Physics Fluid flows may be divided into *incompressible* and *compressible* flows, depending on their characteristic flow speed. Quantifying this speed, a frequently occurring dimensionless parameter is the well-known *Mach number*, defined as the flow speed divided by the speed of sound. If the Mach number is smaller than 0.3, the flow is commonly presumed incompressible, since its density variation is below 5% and compressibility effects can be neglected (Anderson, 1991). For a Mach number greater than 0.3, the flow is considered compressible. Such flows play a central role in many applications, ranging from the turbulent flow over airfoils (Kral, 1998), the design of aircraft (Raymer, 2012), the flow through turbine stages in jet engines (Klapdor, 2011), the fuel injection in Diesel engines (Hill and Ouellette, 1999) to the flow in solar power plant chimneys (von Backström and Gannon, 2000; Schlaich et al., 2005).

Even if the flow speed in the bulk flow is below Mach 1, the flow around convexly shaped airfoils or around the tip of rotating turbine blades can exceed the speed of sound locally. The corresponding flow regimes are denoted the *subsonic* and the *supersonic regime*, respectively. A flow speed of Mach 1 is the limiting case and defines the *sonic conditions*.

In nearly all supersonic flows, a physically complex phenomenon, which is called a *shock wave*, arises. A shock wave is very thin, in the order of a few molecular mean free paths which is around 100 nm for air at standard conditions. It is governed by a sudden, drastic change of several thermodynamic properties of the fluid, such as the density, the pressure, and the temperature (Anderson, 1990, Section 3.6). There is no general closed formula to determine the shape and the position of a shock wave, thus making it challenging to predict.

In the context of supersonic flow, an exemplary active field of research is the sonic boom prediction (National Research Council, 2002; Cohen et al., 2006), which is closely related to the reduction of supersonic noise generated by supersonic jets (Tam and Tanna, 1982; Brehm et al., 2016). The overall noise reduction is a central task in order to enable supersonic commercial aircraft over populated areas besides the reduction of the pollutant emission on the way to green aviation technologies (Suder, 2012).

Numerics The Euler equations are a mathematical model for the description of inviscid compressible flow, building a system of first-order nonlinear partial differential equations (PDEs). Discretizing nonlinear PDEs has been an active field of research for many years. Popular approaches are the Finite Difference Method (FDM), the Finite Volume Method (FVM), and the Finite Element Method (FEM). The comparably young high-order discontinuous Galerkin (DG) method (Reed and Hill, 1973; Cockburn et al., 2000) combines several advantages of the latter two mentioned. Besides its high-order convergence for smooth solutions on structured and unstructured grids, favorable properties of the DG method lie in its conservativity and in its locality. This is because the cell-wise discontinuous polynomial approximations have to be exchanged only between neighboring cells (Hesthaven and Warburton, 2007). This renders the DG method perfectly suitable for local hp-adaptivity, where the grid size and the order of the polynomial approximation can be adjusted independently. Additionally, the locality of the DG method enables a performant parallelization and provides the facility for High-Performance Computing (HPC) applications (Hindenlang et al., 2012; Altmann et al., 2013).

High-order methods can be characterized by a discretization error proportional to $O(h^P)$, where h and P denote a characteristic length scale of the computational grid and the degree of the approximating polynomial, respectively. A survey in the CFD community yielded the result that a 'high-order method' is widely understood as a numerical method of order $P > 2$ (Wang et al., 2013). The core of high-order methods lies in its $O(h^P)$-convergence. As an example, we compare a high-order solution on a coarse grid with a low-order solution on a fine grid, both having the same number of degrees of freedom (DOF). Since the polynomial degree is the key factor appearing as the exponent in $O(h^P)$, the high-order solution achieves a higher accuracy than the low-order one in this case. However, today's industrial applications in the context of fluid flows are still dominated by the low-order FVM. Especially for large three-dimensional transient problems, the high initial computational costs still limit the application of low-order methods for highly resolved applications, since the generation of boundary-fitted grids is elaborate and computationally expensive. Conversely, various works and workshops such as the European initiative *ADIMGA* (Kroll et al., 2010), the subsequent project *IDIHOM* (Kroll et al., 2015) or the solver framework *FLEXI* (Krais et al., 2020) showed the potential of high-order methods in the context of compressible flow, outperforming established low-order ones.

Shock-capturing is one approach to overcome the problem of an oscillating polynomial approximation in the vicinity of discontinuous flow phenomena. This effect is also known as the *Gibb's phenomenon* and arises if a discontinuity does not coincide with the boundary of a grid cell in DG methods. Additionally, it leads to a loss of global high-order accuracy. Shock-capturing strategies circumvent the problem of oscillating numerical solutions by limiting them in a total variation diminishing (TVD)-like manner (van Leer, 1973; Cockburn and Shu, 1989) or aim for a skillful reconstruction of the solution in the affected cells as realized in weighted essentially non-oscillatory (WENO) schemes (Liu et al., 1994; Shu, 2016). An inherent shortcoming of these approaches is, besides the larger implementation effort compared to other approaches of this type, that the convergence to steady-state problems can sometimes hardly be obtained (Barter, 2008). Another sub-class is based on the addition of a second-order *artificial viscosity* term to the governing equations, smoothing the solution over a viscous layer where it can be adequately resolved by the numerical scheme (Von Neumann and Richtmyer, 1950; Bassi and Rebay, 1997; Persson and Peraire, 2006). The simplicity of this approach seems appealing at first glance; however, it fails to preserve the high-order convergence properties of the original method.

Shock-fitting approaches go back to the fundamental work by Moretti and Abbett (1966) who were the first to solve the supersonic blunt-body problem for practical applications. This was accomplished by the use of an FDM with high accuracy. The supersonic blunt-body problem denotes a flow, where a detached, curved shock wave, often denoted a *bow shock*, stands in front of a blunt-body. Shock-fitting approaches have in common that the shock wave coincides with one boundary of the computational domain, across which the flow quantities can be determined by evaluating the Rankine-Hugoniot conditions. We refer to the textbook by Salas (2010) for an introduction and a detailed overview of shock-fitting approaches.

Unfitted methods, *extended* methods, and methods on *fictitious domains* are widely used as synonyms in the literature and describe a prominent class of methods, where the approximation space is *extended* or *enriched* in the area of interest by an additional set of DOF. A sub-class of these methods is the immersed boundary method (IBM), which *embeds* the domain of interest into a simple Cartesian or triangular background grid (Peskin, 1972; Mittal and Iaccarino, 2005). Here, the main effort is shifted from grid generation to incorporating immersed boundaries, which represent, for example, the surface solid bodies or fluid interfaces, into the discretization approach. By contrast, classical approaches rely on *boundary-fitted* domains, where the computational domain coincides with the physical one (Feistauer et al., 2003; Ferziger and Perić, 2008). For large unsteady three-dimensional test cases with complex geometries, the computational costs increase drastically, since boundary-fitted grids have to be regenerated in every time-step. Furthermore, automatically generated boundary-fitted grids usually need to be optimized by hand for complex geometries. This process is completely omitted in an IBM.

1.1 The *BoSSS* Framework

The $\underline{B}$ounded $\underline{S}$upport $\underline{S}$pectral $\underline{S}$olver *(BoSSS)*[1] is a flexible framework for the development, evaluation and application of a numerical discretization of PDEs based on high-order DG methods (Kummer et al., 2020). The *BoSSS* framework makes use of the Message Passing Interface (MPI) and incorporates load balancing strategies, rendering it a capable tool for HPC applications. It features rapid prototyping as well as workflow and data management facilities, for example, for quick testing of new numerical models or the automatic submission of production runs. The *BoSSS* framework is written in the object-oriented C#-programming language, which is part of the *Microsoft .NET* framework.

History   The development of the *BoSSS* framework was initiated at the Chair of Fluid Dynamics (FDY) at the Technical University of Darmstadt in 2008 with the goal of establishing a generally applicable and well-extensible code basis for the development of high-order numerical schemes for challenging physical problems. One year later, it was presented to the public for the first time (Kummer et al., 2009). Ever since, the *BoSSS* framework has been developed into a large library in the context of DG methods. In particular, an extended discontinuous Galerkin (XDG) method for incompressible multi-phase flows (Kummer, 2016) and a discontinuous Galerkin immersed boundary method (DG IBM) for subsonic compressible flows (Müller et al., 2017) as

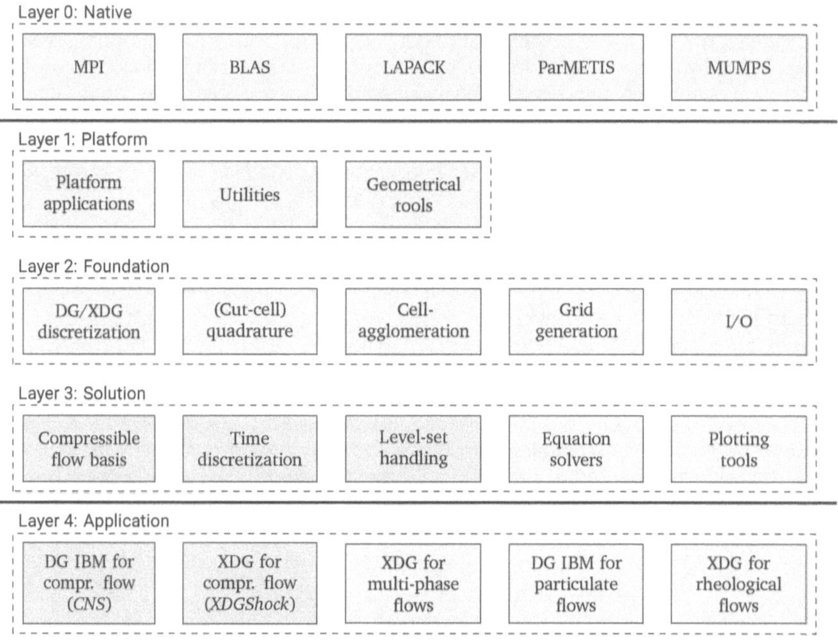

Figure 1.1: Structure of the *Bounded Support Spectral Solver (BoSSS)* framework (Kummer et al., 2020). During the course of this work, existing implementations were refactored and optimized. Furthermore, novel methodologies were added to the highlighted modules (adapted from Müller, 2014, Figure 1.1).

well as for particulate flows (Krause and Kummer, 2017) have been added. Ongoing works deal with rheological flows (Kikker, 2020), combustion, fluid-structure interaction, moving Voronoi grids and further related topics. Figure 1.1 shows the structure of the *BoSSS* framework and the modules which were newly developed and optimized during the course of this work.

Main feature   A main feature of the *BoSSS* framework is its capability to solve steady-state and time-dependent problems, which may additionally contain moving interfaces and time-dependent domains (Kummer et al., 2020). For that, a sharp interface description is applied by means of a level-set function. In this context, advanced techniques for the treatment of moving interfaces are inherently necessary (Utz and Kummer, 2017; Utz et al., 2017; Kummer et al., 2018). Many numerical approaches require the elaborate and, thus, expensive creation of boundary-fitted grids, limiting the applications especially for large three-dimensional problems with moving interfaces or boundaries. Prominent examples are the flow around rotating turbine plates or the injection of fuel in combustion engines. An XDG method in combination with a sharp interface description is a promising tool on the road to a highly accurate numerical solution of these challenging tasks.

Context of this work   This work mainly deals with the highlighted modules in Figure 1.1, which are all related to the simulation of supersonic compressible flow. In particular, we focus on time-efficient *shock-capturing* and *shock-fitting* in the context of a DG IBM and an XDG method, respectively. The reader is referred to the works by Müller (2014), Müller et al. (2017), Krämer-Eis (2017), and Geisenhofer et al. (2019) for related topics, which are beyond the scope of this work.

1.2 Objective and Outline of This Work

The *objective of this work* is to develop a robust and efficient high-order numerical scheme for the simulation of supersonic compressible flow with discontinuous flow phenomena. To achieve this, we employ a discontinuous Galerkin immersed boundary method (DG IBM) and an extended discontinuous Galerkin (XDG) method in the *BoSSS* framework.

The approaches to achieve this objective can be split into the main categories of *shock-capturing* and *shock-fitting*. This work focuses on the numerical treatment of discontinuous flow phenomena by a shock-capturing strategy based on artificial viscosity in a DG IBM. At the same time the computational performance is improved by an adaptive local time-stepping (LTS) scheme. Global high-order accuracy is usually lost when using shock-capturing approaches. This motivates the use of a high-order XDG method, where discontinuous flow phenomena are represented by a sharp interface description. On the road to regaining high-order convergence, we present a novel reconstruction technique of a level-set function in the context of shock-fitting. The zero iso-contour of this level-set function describes the shape and the position of a shock front in a sub-cell accurate manner.

Mathematical and physical fundamentals   In Chapter 2, we introduce the underlying mathematical model by means of a conservative formulation of the Euler equations for inviscid compressible flow. They are supplemented with a material model by using the ideal gas law as an equation of state (EOS) for a calorically perfect gas. Furthermore, the relevant theory of shock and expansion waves is explained.

The spatial discretization   In Chapter 3, the spatial discretization of the employed DG methods is presented. Thereby, we focus on the characteristic features of unfitted methods. In particular, we present the incorporation of a level-set function into the discretization approach. We briefly introduce the concept of cell-agglomeration, which is capable of reducing the maximum stable time-step size in the context of explicit time-integration schemes and additionally improves the conditioning of the system matrix.

Shock-capturing on cut-cells   Shock-capturing strategies prevent undesired oscillations of the polynomial approximation in the vicinity of discontinuous flow phenomena. In this work, we employ a shock-capturing strategy based on artificial viscosity. We improve the efficiency of this strategy by means of an adaptive LTS scheme.

In Chapter 4, we first address the topic of explicit time-integration schemes, before we present a novel adaptive LTS scheme in the context of a DG IBM for compressible flow. Small and ill-shaped cut-cells may be removed from the computational grid by applying a cell-agglomeration technique. The remaining cut-cells still place a severe restriction on the maximum stable time-step size in the context of explicit time-integration schemes, which are widely used for advancing the Euler equations in time due to their hyperbolic character. As a remedy, we present an adaptive LTS scheme which groups the cells in clusters according to their cell-local time-step size. The numerical solution in each cell cluster is integrated in time separately. For unsteady flows, an adaptive rebuild of the cell clustering is inherently necessary to improve the computational performance.

In Chapter 5, we present a two-step shock-capturing strategy, which consists of a modal-decay indicator and a smoothing procedure based on artificial viscosity. Thereby, an additional second-order diffusive term is added to the governing equations so that discontinuous flow phenomena are spread over a layer where they can be adequately resolved by the numerical scheme. The two-step shock-capturing strategy and the adaptive LTS scheme are extended for the use on an agglomerated cut-cell grid. To the best of our knowledge, we were the first to employ such an approach in the context of a DG IBM (Geisenhofer et al., 2019). Hence, this can be considered one of the main contributions of this work. We obtain our numerical results by employing the *Compressible Navier-Stokes (CNS)* solver, which is based on the works by Müller (2014), Müller et al. (2017), Krämer-Eis (2017), and Geisenhofer et al. (2019) to a large extent, and compare them to classical benchmarks from the literature.

Shock-fitting   In Chapter 6, we derive a novel technique for the sub-cell accurate reconstruction of a shock front, which is described by the zero iso-contour of a level-set function. The presented proof of concept for a pseudo-two-dimensional stationary normal shock wave can be considered another main contribution of this work. The reconstruction algorithm is based on an implicit pseudo time-stepping procedure, which corrects the position of the shock interface inside a cut background cell, before the entire flow field is advanced in time. This work builds a fundamental basis for a high-order XDG method for supersonic compressible flow.

Conclusion   In Chapter 7, we summarize the main aspects of this work and give an outlook for future work. We show first promising results of a two-dimensional supersonic blunt-body problem. The results clearly indicate the need of advanced implicit time-integration schemes for a stable and robust XDG method, when extending the reconstruction procedure of the shock front to the higher-dimensional case.

Note on the way of citing   Nearly all written works in research, such as papers, books, reports and theses, are built on the works of others, as is this work. We refer to the works of others by using the *authoryear*-style by writing *Last name et al. (2020)* inside a sentence and *(Last name et al., 2020)* in parenthetical style. If a larger part of a chapter or section is based on the work of others, we indicate this in the beginning explicitly. Furthermore, we refer to text taken from our own publications by adding a footnote to the heading or to the end of a paragraph. All relevant references of this work can be found in the Bibliography.

2 Supersonic Compressible Flow

Compressible flows appear in many industrial applications such as in jet engines or in commercial aircraft. Due to the large variation of the density in such flows, the well-known incompressibility constraint, which considers fluid particles of constant density, loses its validity as a physical and mathematical simplification of the Navier-Stokes equations. High-speed compressible flows are dominated by inertia forces so that the effects of viscosity and heat transfer can be neglected. In the literature, the Euler equations for inviscid compressible flow are often applied as the underlying model.

An active field of research is the sonic boom prediction, since it prevents commercial and cargo supersonic flight over land and populated ares due to shock-associated noise (Barter, 2008). In the beginning of the millennium, the United States National Research Council (2002) assigned the task to investigate the feasibility of quite supersonic aircraft to the National Aeronautics and Space Administration (NASA). The sonic boom mitigation was determined as the key barrier to break through. Consequently, a deep-insight physical knowledge of supersonic flow, which contains several macroscopically discontinuous flow features such as shock waves, is essential in order to derive recommendations for the design of airplanes. Furthermore, suitable Computational Fluid Dynamics (CFD) techniques with high accuracy are needed to support, for example, experimental wind-tunnel testing during the development phase. High-order discontinuous Galerkin (DG) methods can take on the task of simulating realistic flow scenarios with high Mach numbers, for which we present two different numerical approaches in Chapters 5 and 6, respectively.

This chapter is structured as follows: We deal with inviscid compressible flow in Section 2.1 and give an overview of shock wave phenomena in Section 2.2. The first part includes a conservative formulation of the Euler equations in Section 2.1.1, remarks about the equation of state (EOS) in Section 2.1.2, and notes about the applied non-dimensionalisation procedure in Section 2.1.3. The second part gives a tailored introduction to the mathematical relations across normal and oblique shock waves in Sections 2.2.1 and 2.2.2, respectively, and to the basic theory of unsteady wave motion in Section 2.2.3. The chapter closes with a brief discussion of shock wave reflections in Section 2.2.4.

Compressibility and flow regimes In contrast to an incompressible flow where the density ρ is assumed to be constant ($D\rho/Dt = 0$), a compressible (real) flow inherently features a variable density. Therefore, the concept of compressibility and the resulting flow regimes are briefly sketched in the following paragraph (based on Anderson, 1990, Section 1.2 and 8.5).

We consider a fluid element of unit mass with a specific volume v and the corresponding density $\rho = 1/v$. In a flow, this fluid element experiences a pressure p which is caused by its neighboring fluid elements. If the pressure is increased by an infinitesimal amount $\mathrm{d}p$, the fluid element is compressed by $\mathrm{d}v$. This leads to the definition of the compressibility

$$\tau = -\frac{1}{v}\frac{\mathrm{d}v}{\mathrm{d}p}.$$
(2.1)

In general, gases have a much larger compressibility than liquids. For example, the isothermal compressibility is $\tau_T = 5 \cdot 10^{-10}\,\mathrm{m^2/N}$ for water and $\tau_T = 5 \cdot 10^{-5}\,\mathrm{m^2/N}$ for air at sea level conditions. In most technical applications it is assumed that compressible effects are negligible for gas flows with a speed less than 0.3 of the speed of sound a, since the density variation is low. One of the fundamental quantities for the characterization of compressible flow is the *Mach number*

$$\mathrm{M} = \frac{|\mathbf{u}|}{a},$$
(2.2)

where $|\mathbf{u}|$ is the L^2-norm of the flow velocity vector, see Definition 3.1 and Section 2.1.3 for details. Depending on the Mach number we distinguish three different flow regimes

$$\mathrm{M} \begin{cases} < 1: & \text{subsonic flow}, \\ = 1: & \text{sonic flow}, \\ > 1: & \text{supersonic flow}. \end{cases}$$
(2.3a)

These flow regimes differ considerably in the occurring physical phenomena. For example, shock waves arise in supersonic flows as presented in Section 2.2.

2.1 The Euler Equations

The Euler equations are a physical and mathematical model for inviscid compressible flow and can be derived from the Navier-Stokes equations for large Reynolds numbers

$$\mathrm{Re} = \frac{\rho_\infty u_\infty l_\infty}{\mu_\infty} \to \infty,$$
(2.4)

where ρ_∞, u_∞, l_∞, and μ_∞ denote characteristic, usually free-stream, values of the density, the velocity, the length scale, and the dynamic viscosity, respectively. A detailed derivation can be found, for example, in the textbooks by Anderson (1990) and Spurk and Aksel (2010).

In the following, we introduce a conservative form of the Euler equations with dimensions and a dimensionless form in Sections 2.1.1 and 2.1.3, respectively. We present the applied EOS in Section 2.1.2. The notation is based on the works by Müller (2014), Krämer-Eis (2017), and Geisenhofer et al. (2019) for consistency.

2.1.1 Conservative Form With Dimensions

We consider the two-dimensional Euler equations, which consist of the conservation laws for mass, momentum, and energy, in a differential conservative form

$$\frac{\partial \mathbf{U}}{\partial t} + \frac{\partial \mathbf{F}_1(\mathbf{U})}{\partial x_1} + \frac{\partial \mathbf{F}_2(\mathbf{U})}{\partial x_2} = 0.$$
(2.5)

Here, $\mathbf{U} \in \mathbb{R}^4$ is the state vector of conserved quantities

$$\mathbf{U} = \begin{pmatrix} \rho \\ \rho u_1 \\ \rho u_2 \\ \rho E \end{pmatrix}, \tag{2.6}$$

and $\mathbf{F}_1(\mathbf{U}) : \mathbb{R}^4 \to \mathbb{R}^4$ and $\mathbf{F}_2(\mathbf{U}) : \mathbb{R}^4 \to \mathbb{R}^4$ are the convective flux vectors

$$\mathbf{F}_1(\mathbf{U}) = \begin{pmatrix} \rho u_1 \\ \rho u_1 u_1 + p \\ \rho u_1 u_2 \\ u_1(\rho E + p) \end{pmatrix}, \qquad \mathbf{F}_2(\mathbf{U}) = \begin{pmatrix} \rho u_2 \\ \rho u_1 u_2 \\ \rho u_2 u_2 + p \\ u_2(\rho E + p) \end{pmatrix}. \tag{2.7}$$

In Equations (2.5) to (2.7), $\mathbf{x} = (x_1, x_2)^\mathsf{T} \in \mathbb{R}^2$ is the spatial coordinate vector, $t \in \mathbb{R}^+$ is the time, $\rho \in \mathbb{R}^+$ is the fluid density, $\mathbf{m} = (\rho u_1, \rho u_2)^\mathsf{T} \in \mathbb{R}^2$ is the momentum vector, $\mathbf{u} = (u_1, u_2)^\mathsf{T} \in \mathbb{R}^2$ is the velocity vector, $\rho E \in \mathbb{R}^+$ is the total energy, and $p \in \mathbb{R}^+$ is the pressure. The total energy is the sum of the internal energy $\rho e \in \mathbb{R}^+$ and the kinetic energy so that

$$\underbrace{\rho E}_{\text{total energy}} = \underbrace{\rho e}_{\text{internal energy}} + \underbrace{\frac{1}{2}\rho\, \mathbf{u} \cdot \mathbf{u}}_{\text{kinetic energy}}. \tag{2.8}$$

The specific total energy is denoted by $E \in \mathbb{R}^+$, the specific internal energy by $e \in \mathbb{R}^+$, and the specific enthalpy by $h = e + p/\rho \in \mathbb{R}^+$. *Specific* quantities are defined per unit mass. These expressions are usually used as synonyms in the literature.

Equation system (2.5) is not closed yet, since it misses an EOS for the pressure $p = p(\rho, e)$, which we introduce in the next section.

2.1.2 Equation of State

We restrict ourselves to flow configurations where the *perfect gas assumption* is valid. Here, intermolecular forces, such as the Van der Waals forces, are neglected. This assumption is suitable for a wide range of research and engineering applications (Anderson, 1990, Section 1.4.1). The perfect gas assumption is not valid at low temperatures and high pressures, since there intermolecular forces become important, as well as in hypersonic flows. It also breaks down in reactive flows, since chemical reactions lead to a variation of the gas constant. The *ideal gas law*

$$p = \rho R_{\text{gas}} T \tag{2.9}$$

holds for a perfect gas. In Equation (2.9), $R_{\text{gas}} \in \mathbb{R}^+$ is the specific gas constant and $T \in \mathbb{R}^+$ is the temperature. We use this EOS in the remainder of this work.

A brief excursion into thermodynamics The following paragraph briefly introduces several thermodynamic principles which are relevant for this work (based on Anderson, 1990, Section 1.4). In a *thermally perfect gas*, intermolecular forces are neglected and no chemical reactions take place so that the internal energy e and the enthalpy $\hat{h}$ are functions of solely the temperature. This yields the expressions

$$e = e(T), \qquad \hat{h} = \hat{h}(T), \tag{2.10a}$$

$$\mathrm{d}e = c_v(T)\,\mathrm{d}T, \qquad \mathrm{d}\hat{h} = c_p(T)\,\mathrm{d}T, \tag{2.10b}$$

where $c_v \in \mathbb{R}^+$ and $c_p \in \mathbb{R}^+$ are the specific heats at constant volume and pressure. We assume $e(T = 0) = \hat{h}(T = 0) = 0$ and the specific heats being constant, which results in the system for a *calorically perfect gas*

$$e = c_v T, \qquad \hat{h} = c_p T, \tag{2.11}$$

being valid at air temperatures $T \lesssim 1000\,\mathrm{K}$. We apply this assumption in the remainder of this work. We define the heat capacity ratio $\gamma \in \mathbb{R}^+$ as

$$\gamma = \frac{c_p}{c_v} = \frac{\hat{h}}{e}, \tag{2.12}$$

where $\gamma = 1.4$ is a suitable assumption for diatomic gases such as air at standard conditions. Taking the relation

$$c_p - c_v = R_{\mathrm{gas}} \tag{2.13}$$

into account, the ideal gas law (2.9) can also be written as

$$p = \rho R_{\mathrm{gas}} T = (\gamma - 1)\rho e. \tag{2.14}$$

Müller (2014) discusses different EOS and their implementation in the *Compressible Navier-Stokes (CNS)* solver of the *BoSSS* framework. An example is the stiffened gas law which extends the ideal gas law (2.9). It adds an additional term which models the stiffness of the fluid with a pre-loaded pressure term.

The *first law of thermodynamics*

$$\mathrm{d}e = \delta \hat{Q} + \delta W \tag{2.15}$$

states that the change of internal energy $\mathrm{d}e$ in a closed stationary system is equal to the sum of the heat $\delta\hat{Q}$ added through the system boundary and the work δW done on the system by its surrounding. We define an *isentropic process* as a process where no heat is added to the system or taken away from the system (adiabatic process), and no dissipative phenomena, such as friction, mass diffusion or thermal conduction, are present (reversible process).

The *second law of thermodynamics* states in which direction a process takes place. It also introduces the concept of entropy

$$\mathrm{d}\hat{s} = \underbrace{\frac{\delta\hat{Q}}{T}}_{=0,\ \text{if adiabatic}} + \underbrace{\mathrm{d}\hat{s}_{\mathrm{irrev}}}_{=0,\ \text{if reversible}}, \tag{2.16}$$

where the change of the entropy $\mathrm{d}\hat{s}$ in a system consists of the sum of the heat added/dissipated to/from the system and of irreversible dissipative phenomena occurring within the system. These phenomena are associated with an irreversible change of entropy $\mathrm{d}\hat{s}_{\mathrm{irrev}}$. The *second law of thermodynamics* states that entropy can never decrease so that

$$\mathrm{d}\hat{s}_{\mathrm{irrev}} \geq 0 \tag{2.17}$$

in a closed system. Finally, we state the expressions

$$\hat{s}_2 - \hat{s}_1 = c_p \ln\left(\frac{T_2}{T_1}\right) - R_{\mathrm{gas}} \ln\left(\frac{p_2}{p_1}\right), \tag{2.18a}$$

$$\hat{s}_2 - \hat{s}_1 = c_v \ln\left(\frac{T_2}{T_1}\right) + R_{\mathrm{gas}} \ln\left(\frac{v_2}{v_1}\right), \tag{2.18b}$$

which allow the calculation of the change of entropy between two states for a calorically perfect gas. Here, the entropy $\hat{s}$ is always a function of $\hat{s} = \hat{s}(T, p)$ or $\hat{s} = \hat{s}(T, v)$, respectively (Anderson, 1990, Section 1.4.5).

An isentropic process is a process where the change of entropy is $\mathrm{d}\hat{s} = 0$ along the path of a fluid element. This implies that the process is adiabatic and reversible, see Equation (2.16). Frequently in the analysis of compressible flow, an isentropic process is described by the *isentropic relations*

$$\frac{p_2}{p_1} = \left(\frac{\rho_2}{\rho_1}\right)^{\gamma} = \left(\frac{T_2}{T_1}\right)^{\gamma/(\gamma-1)}. \tag{2.19}$$

The idea of an isentropic process seems to be too idealized to be of practical relevance at first glance. However, the flow field can be considered isentropic in many practical applications when boundary layer effects are neglected. This is a suitable assumption for high Mach number flows where the boundary layers are usually very thin.

2.1.3 Non-Dimensional Conservative Form and Quantities

It is common to derive and apply a non-dimensional form of an equation in order to transfer results from small to large scales, or vice versa, in experiments and numerics. In order to obtain the dimensionless parameters for inviscid compressible flow and a non-dimensional form of the Euler equations (2.5), we first introduce the non-dimensional independent variables

$$\mathbf{x}' = \frac{\mathbf{x}}{l_\infty}, \qquad t' = \frac{u_\infty t}{l_\infty}, \tag{2.20}$$

as well as the non-dimensional dependent variables

$$\rho' = \frac{\rho}{\rho_\infty}, \qquad \mathbf{m}' = \frac{\mathbf{m}}{\rho_\infty u_\infty}, \qquad (\rho E)' = \frac{\rho E}{\rho_\infty u_\infty^2},$$

$$p' = \frac{p}{p_\infty}, \qquad \mathbf{u}' = \frac{\mathbf{u}}{u_\infty}, \qquad e' = \frac{e}{u_\infty^2}, \qquad T' = \frac{\rho_\infty R_{\mathrm{gas}} T}{p_\infty}, \tag{2.21}$$

where $(\cdot)'$ denotes the non-dimensional quantities and $(\cdot)_\infty$ denotes the reference values. Note that the introduction of a reference pressure p_∞ is necessary in order to express the reference Mach number M_∞ independently of any other non-dimensional parameter (Krämer-Eis, 2017). Subsequently, we follow the approach presented by Müller (2014) and Krämer-Eis (2017). Inserting Equations (2.20) and (2.21) into the Euler equations (2.5) leads to the non-dimensional convective flux vectors

$$\mathbf{F}_1(\mathbf{U}') = \frac{1}{\gamma M_\infty^2} \begin{pmatrix} (\rho u_1)' \\ (\rho u_1)'\, u_1' + p' \\ (\rho u_1)'\, u_2' \\ u_1' \left((\rho E)' + p'\right) \end{pmatrix}, \qquad \mathbf{F}_2(\mathbf{U}') = \frac{1}{\gamma M_\infty^2} \begin{pmatrix} (\rho u_2)' \\ (\rho u_1)'\, u_2' \\ (\rho u_2)'\, u_2' + p' \\ u_2' \left((\rho E)' + p'\right) \end{pmatrix}, \qquad (2.22)$$

where γ and $M_\infty = |\mathbf{u}_\infty| / a_\infty$ are the non-dimensional parameters for inviscid compressible flow. Without loss of generality, we set $M_\infty = 1/\sqrt{\gamma}$. With this specific choice, the non-dimensional convective fluxes (2.22) have the same form as their corresponding version with dimensions (2.7) (Feistauer et al., 2003). Consequently, the initial conditions of any test case can be implemented by using the quantities with dimensions without any further adaption. We use the relation $\rho_\infty u_\infty^2 / p_\infty = \gamma M_\infty^2 = 1$ and obtain the non-dimensional form of the ideal gas law after inserting Equations (2.20) and (2.21) into the ideal gas law (2.9)

$$p' = \rho' T' = (\gamma - 1)\rho' e' \,. \qquad (2.23)$$

From now on, we omit the prime for better readability when using the non-dimensional form.

Speed of sound and local Mach number   A sound wave is propagating through a medium at the speed of sound a while slightly changing the physical quantities of the surrounding medium. We follow the literature and denote such a wave a *weak wave* (Anderson, 1990, Section 3.3). This process can be assumed to be isentropic, which leads to the definition of the speed of sound

$$a = \sqrt{\left(\frac{\partial p}{\partial \rho}\right)_s} = \frac{v}{\tau_s} \,. \qquad (2.24)$$

This equation directly relates the isentropic compressibility $\tau_s = (-1/v\, \mathrm{d}v/p)_s$ of the medium to the speed of sound a, see also Equation (2.1). The speed of sound is $a = 340.9\,\mathrm{m/s}$ for standard air conditions (Anderson, 1990, Section 3.3).

The isentropic relations (2.19) can be reformulated, yielding the expressions

$$pv^\gamma = \text{const.} \qquad (2.25)$$

After combining Equations (2.19), (2.24) and (2.25) we obtain the speed of sound

$$a = \sqrt{\gamma \frac{p}{\rho}} \qquad (2.26)$$

for a calorically perfect gas. Inserting the ideal gas law (2.9) into Equation (2.26) yields

$$a = \sqrt{\gamma R_{\text{gas}} T} \,. \qquad (2.27)$$

In general, the Mach number is a local variable in the flow field. Using $M_\infty = 1/\sqrt{\gamma}$ Equations (2.2) and (2.27) leads to the definition of the *local Mach number*

$$M_l = \frac{|\mathbf{u}|}{a} = \frac{|\mathbf{u}|}{\sqrt{\frac{p}{\rho}}} M_\infty = \frac{|\mathbf{u}|}{\sqrt{\gamma \frac{p}{\rho}}} \,. \qquad (2.28)$$

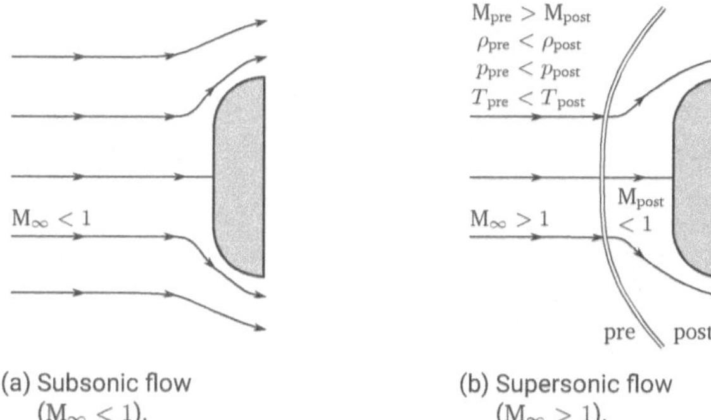

(a) Subsonic flow
($M_\infty < 1$).

(b) Supersonic flow
($M_\infty > 1$).

Figure 2.1: Streamlines around a blunt body for a subsonic and a supersonic flow. If $M_\infty < 1$, the streamlines change continuously due to the presence of the body. If $M_\infty > 1$, a detached, oblique shock wave forms in front of the body. Directly behind the shock, the flow is subsonic ($M_{post} < 1$) (adapted from Anderson, 1990, Figure 3.8).

2.2 Shock and Expansion Waves

This section introduces the basic theory of shock and expansion waves. Figure 2.1 shows the flow over a blunt body for a subsonic and a supersonic flow. In all flows, information propagates with the speed of sound via molecular collisions. We consider a sound wave which is emanated from the blunt body due to the molecular collisions. In a supersonic flow, the sound wave can no longer move upstream and forewarn the upstream flow of the presence of the body. This causes all sound waves, which are emanate from the body, to coalesce close in front of it, see Figure 2.1b. As a result, a detached, oblique shock wave forms, directly behind which the flow becomes subsonic ($M_{post} < 1$). Further downstream, the flow may become supersonic again, since it is accelerated along the convex body surface. Across a shock wave, the physical quantities show a macroscopically discontinuous behavior. In nature, a shock wave is very thin, around 100 nm for air at standard conditions (Anderson, 1990, Section 3.6).

Subsequently, we introduce the relations to calculate the change of the flow quantities across a stationary normal shock wave in Section 2.2.1. We extend these relations for two-dimensional stationary oblique shock waves in Section 2.2.2. Moving shock waves are considered in Section 2.2.3, where the frame of reference is no longer fixed to the shock wave. In Section 2.2.4, we briefly introduce shock wave reflection phenomena. The following sections are based on Anderson (1990, Chapters 3, 4, and 7) and Ben-Dor (2007, Chapter 1).

2.2.1 Normal Shock Relations

In Figure 2.1b, the stationary shock wave can be considered as a normal shock wave in the near of the centerline. A normal shock wave is a straight, non-curved flow phenomenon across which mass, momentum, and energy are conserved. For that, we fix the reference frame to the shock wave and use the jump operator $[\![\cdot]\!]$, which we will introduce in Definition 3.3, for stating the jump conditions as

$$\llbracket \rho \mathbf{u} \cdot \mathbf{n} \rrbracket = 0 \quad \text{(continuity equation)}, \tag{2.29a}$$

$$\llbracket \rho \mathbf{u}\mathbf{u} \cdot \mathbf{n} + p\mathbf{n} \rrbracket = 0 \quad \text{(momentum equation)}, \tag{2.29b}$$

$$\llbracket (\rho E + p)\,\mathbf{u} \cdot \mathbf{n} \rrbracket = \llbracket (\rho e + 1/2\,\rho\,\mathbf{u} \cdot \mathbf{u} + p)\,\mathbf{u} \cdot \mathbf{n} \rrbracket = 0 \quad \text{(energy equation)}, \tag{2.29c}$$

where $\mathbf{n}$ is the normal vector of the shock wave. Note that the tangential component of the velocity vector is conserved across a shock wave. In the case of a normal shock wave with $\mathbf{n} = (1,0)^{\mathsf{T}}$, $\llbracket u_2 \rrbracket = 0$ holds. Evaluating the jump conditions (2.29) or the Euler equations (2.5) in a rectangular control volume across a normal shock wave, respectively, results in the simple one-dimensional equations

$$\rho_{\text{pre}} u_{1,\text{pre}} = \rho_{\text{post}} u_{1,\text{post}} \quad \text{(continuity equation)}, \tag{2.30a}$$

$$p_{\text{pre}} + \rho_{\text{pre}} u_{1,\text{pre}}^2 = p_{\text{post}} + \rho_{\text{post}} u_{1,\text{post}}^2 \quad \text{(momentum equation)}, \tag{2.30b}$$

$$\hat{h}_{\text{pre}} + \frac{u_{1,\text{pre}}^2}{2} = \hat{h}_{\text{post}} + \frac{u_{1,\text{post}}^2}{2} \quad \text{(energy equation)}, \tag{2.30c}$$

where $(\cdot)_{\text{pre}}$ denotes the pre-shock quantities and $(\cdot)_{\text{post}}$ denotes the post-shock quantities in a steady flow. In the case of a calorically perfect gas, we can add the relations

$$p = \rho R_{\text{gas}} T, \tag{2.9, repeated}$$

$$\hat{h} = c_p T, \tag{2.11, repeated}$$

which in total leads to an equation system with five unknowns ρ_{post}, $u_{1,\text{post}}$, $\hat{h}_{\text{post}}$, p_{post}, and T_{post} and five equations. Skipping some algebraic manipulations, we obtain the following expression for the Mach number M_{post} behind a normal shock wave

$$\text{M}_{\text{post}}(\text{M}_{\text{pre}}) = \sqrt{\frac{1 + [(\gamma - 1)/2]\,\text{M}_{\text{pre}}^2}{\gamma \text{M}_{\text{pre}}^2 - (\gamma - 1)/2}}. \tag{2.31}$$

The post-shock Mach number (2.31) is a function of solely the pre-shock Mach number. The limiting case of a shock wave is given by $\text{M}_{\text{pre}} = \text{M}_{\text{post}} = 1$. This case is defined as an infinitely weak shock wave, which is also called a *Mach wave*. The relations between pre- and post-shock values of the other relevant quantities are listed below, yielding

$$\frac{\rho_{\text{post}}}{\rho_{\text{pre}}} = \frac{u_{1,\text{pre}}}{u_{1,\text{post}}} = \frac{(\gamma + 1)\text{M}_{\text{pre}}^2}{2 + (\gamma - 1)\text{M}_{\text{pre}}^2}, \tag{2.32a}$$

$$\frac{p_{\text{post}}}{p_{\text{pre}}} = 1 + \frac{2\gamma}{\gamma + 1}\left(\text{M}_{\text{pre}}^2 - 1\right), \tag{2.32b}$$

$$\frac{T_{\text{post}}}{T_{\text{pre}}} = \frac{\hat{h}_{\text{post}}}{\hat{h}_{\text{pre}}} = \left[1 + \frac{2\gamma}{\gamma + 1}\left(\text{M}_{\text{pre}}^2 - 1\right)\right]\left[\frac{2 + (\gamma - 1)\text{M}_{\text{pre}}^2}{(\gamma + 1)\text{M}_{\text{pre}}^2}\right]. \tag{2.32c}$$

Equations (2.32a) to (2.32c) are also solely functions of M_{pre}. Across normal shock waves, the density, the pressure, the enthalpy, the energy, and the temperature increase, whereas the velocity and the local Mach number decrease, see also Figure 2.1b. By contrast, the momentum stays constant.

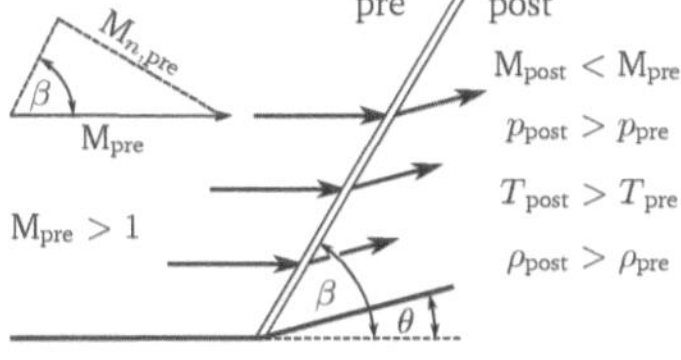

(a) Concave corner (oblique shock wave).

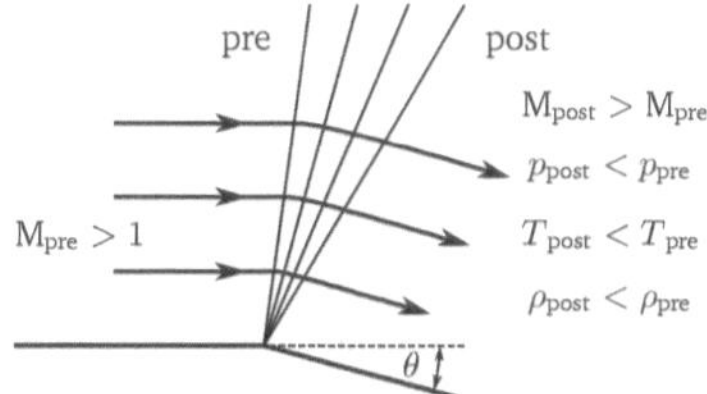

(b) Convex corner (expansion fan).

Figure 2.2: Supersonic flow over corners. The flow quantities are discontinuous across the shock at the concave corner. The streamlines are always parallel to the wall. By contrast, the flow quantities change continuously over the expansion fan at the convex corner. The streamlines are parallel to the wall ahead and behind the expansion fan (adapted from Anderson, 1990, Figure 4.4).

2.2.2 Oblique Shock Relations

Figure 2.2 schematically sketches a supersonic flow with $M_{pre} > 1$ over a concave and over a convex corner, which are both inclined at an angle of $\pm\theta$.

At the concave corner in Figure 2.2a, the streamlines are deflected towards the bulk flow, being uniform and parallel to the wall in the pre- and post-shock state. Additionally, an oblique shock wave forms at the beginning of the ramp and encloses an angle β with the horizontal axis. The velocity vector can be decomposed into a state normal to the oblique shock and a state parallel to the oblique shock. We skip the calculation of the parallel state at this point, since it is not relevant for the remainder of this work. Note that the oblique shock wave becomes detached and curved, standing in front of the ramp, when keeping $M_{pre} = \mathrm{const.} > 1$ and increasing the inclination angle over a critical limit ($\theta > \theta_{max}$). From Figure 2.2a, we can extract the geometrical relation

$$M_{n,pre} = \frac{|\mathbf{u}_{pre}|}{a_{pre}} \sin(\beta) = M_{pre} \sin(\beta) \,, \tag{2.33}$$

where $M_{n,pre}$ is the component of the pre-shock Mach number pointing in the direction of the shock normal. Using Equation (2.33) and further geometrical relations from Figure 2.2a, we can calculate the post-shock Mach number to

$$M_{post} = \frac{M_{n,post}}{\sin(\beta - \theta)} \,, \tag{2.34}$$

with

$$M_{n,post} = \sqrt{\frac{M_{n,pre}^2 + [2/(\gamma - 1)]}{[2\gamma/(\gamma - 1)]M_{n,pre}^2 - 1}} \,. \tag{2.35}$$

As stated in Section 2.2.1, the changes across a normal shock wave are solely governed by M_{pre}, see Equation (2.31). For an oblique shock wave, the changes are governed by M_{pre} and the shock angle β. They can be calculated by using $M_{n,pre}$ instead of M_{pre} in Equation (2.30).

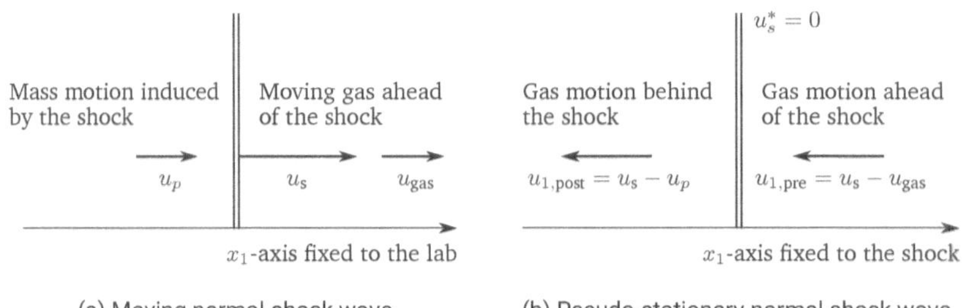

(a) Moving normal shock wave.   (b) Pseudo-stationary normal shock wave.

Figure 2.3: Unsteady wave motion. (a) A normal shock wave with a velocity u_s moves into a gas with a velocity u_{gas}. It induces a mass motion u_p behind it. (b) The reference frame is now fixed to the shock wave so that it appears stationary in space ($u_s^* = 0$). Thus, the normal shock relations (2.30) of the steady case can be reused (adapted from Ben-Dor, 2007, Figure 3.1).

Consequently, normal shock waves are a special case of oblique shock waves for an deflection angle $\beta = \pi/2$.

At the convex corner in Figure 2.2b, the streamlines are deflected away from the bulk flow. An expansion fan develops centered at the corner, where the physical quantities change continuously. Before and after the expansion fan, the flow is uniform and parallel to the wall as it is at the concave corner. Here, the change of all physical quantities across the shock behaves exactly opposite to the concave corner scenario.

2.2.3 Unsteady Wave Motion

In Sections 2.2.1 and 2.2.2, we have considered stationary shock waves with a velocity $u_s = 0$. In the following, we introduce the basics of unsteady wave motion for shock velocities $u_s \neq 0$.

Moving normal shock waves   We schematically sketch a moving normal shock wave and its pseudo-stationary equivalent in Figure 2.3. In Figure 2.3a, the reference frame is fixed to the laboratory, whereas in Figure 2.3b, it is fixed to the shock wave. Consequently, we can reuse the previously introduced theory of stationary shock waves in the latter case.

In Figure 2.3a, the shock wave moves with a velocity u_s to the right into a gas with a velocity u_{gas}. Furthermore, the moving shock wave induces a mass motion with a velocity of u_p behind it. Subsequently, we exploit the Galilean coordinate transformation so that the moving shock wave appears stationary in space (Ben-Dor, 2007, Chapter 3), as shown in Figure 2.3b. Since the reference frame is fixed to the shock wave, these flows are often called *pseudo-stationary* or *pseudo-steady flows*. We obtain the flow velocities $u_{1,\text{pre}}$ and $u_{1,\text{post}}$ ahead and behind the shock in Figure 2.3b by superimposing a velocity equal to the shock velocity u_s but in the opposite direction. This scenario is analogues to the steady cases, as depicted in Figures 2.1b and 2.2a.

We reuse the normal shock relations (2.29) and (2.30) of the stationary case and formulate the shock relations for the unsteady or pseudo-stationary case as

$$[\![\rho(\mathbf{u}_s - \mathbf{w}) \cdot \mathbf{n}]\!] = 0, \tag{2.36a}$$

$$[\![\rho(\mathbf{u}_s - \mathbf{w})(\mathbf{u}_s - \mathbf{w}) \cdot \mathbf{n} + p\mathbf{n}]\!] = 0, \tag{2.36b}$$

$$[\![(\rho E + p)(\mathbf{u}_s - \mathbf{w})]\!] =$$
$$[\![(\rho e + 1/2\,\rho\,(\mathbf{u}_s - \mathbf{w}) \cdot (\mathbf{u}_s - \mathbf{w}) + p)(\mathbf{u}_s - \mathbf{w}) \cdot \mathbf{n}]\!] = 0, \tag{2.36c}$$

where $\mathbf{n}$ is the normal vector of the shock wave and $\mathbf{w}$ is either the velocity vector of the gas ahead of the shock wave $\mathbf{u}_{\text{gas}}$ or the induced mass motion $\mathbf{u}_p$ behind it. Alternatively, Equation (2.36) can be written as

$$\rho_{\text{pre}}(u_s - u_{\text{gas}}) = \rho_{\text{post}}(u_s - u_p), \tag{2.37a}$$

$$p_{\text{pre}} + \rho_{\text{pre}}(u_s - u_{\text{gas}})^2 = p_{\text{post}} + \rho_{\text{post}}(u_s - u_p)^2, \tag{2.37b}$$

$$\hat{h}_{\text{pre}} + \frac{(u_s - u_{\text{gas}})^2}{2} = \hat{h}_{\text{post}} + \frac{(u_s - u_p)^2}{2}. \tag{2.37c}$$

Rearranging Equation (2.37) leads to the *Hugoniot equation*

$$e_{\text{post}} - e_{\text{pre}} = \frac{p_{\text{pre}} + p_{\text{post}}}{2}(v_{\text{pre}} - v_{\text{post}}), \tag{2.38}$$

which is the same like in the stationary case and relates the changes of the thermodynamic quantities across a normal shock wave. By considering a calorically perfect gas with the expressions $e = c_v T$ and $v = R_{\text{gas}} T/p$, we obtain the ratios

$$\frac{T_{\text{post}}}{T_{\text{pre}}} = \frac{p_{\text{post}}}{p_{\text{pre}}} \left(\frac{\frac{\gamma+1}{\gamma-1} + \frac{p_{\text{post}}}{p_{\text{pre}}}}{1 + \frac{\gamma+1}{\gamma-1}\frac{p_{\text{post}}}{p_{\text{pre}}}} \right) \tag{2.39}$$

and

$$\frac{\rho_{\text{post}}}{\rho_{\text{pre}}} = \frac{1 + \frac{\gamma+1}{\gamma-1}\frac{p_{\text{post}}}{p_{\text{pre}}}}{\frac{\gamma+1}{\gamma-1} + \frac{p_{\text{post}}}{p_{\text{pre}}}}. \tag{2.40}$$

In Equations (2.39) and (2.40), the temperature and density ratio across a normal moving shock wave are governed by the pressure ratio $p_{\text{post}}/p_{\text{pre}}$, similar to the stationary case in Equations (2.31) and (2.32), where M_{pre} is the governing parameter. We define the moving shock Mach number as

$$M_s = \frac{u_s}{a_{\text{pre}}} \tag{2.41}$$

and by using Equation (2.32b), the shock velocity reads

$$u_s = a_{\text{pre}} \sqrt{\frac{\gamma+1}{2\gamma}\left(\frac{p_{\text{post}}}{p_{\text{pre}}} - 1\right) + 1}. \tag{2.42}$$

Equation (2.42) states that the shock wave velocity u_s is related to the pressure ratio $p_{\text{post}}/p_{\text{pre}}$ across the shock wave and to the speed of sound a_{pre} upstream of the shock wave. The velocity

$$u_p = \frac{a_{\text{pre}}}{\gamma}\left(\frac{p_{\text{post}}}{p_{\text{pre}}} - 1\right)\sqrt{\frac{\frac{2\gamma}{\gamma+1}}{\frac{p_{\text{post}}}{p_{\text{pre}}} + \frac{\gamma-1}{\gamma+1}}} \tag{2.43}$$

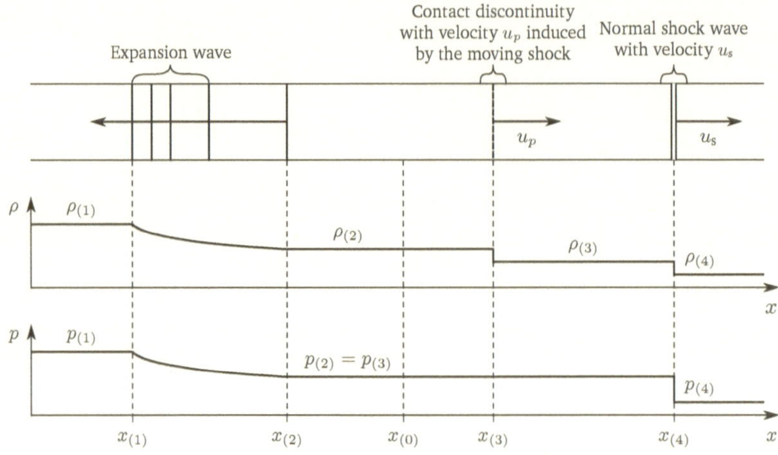

Figure 2.4: Flow configuration in a shock tube after the diaphragm is broken. We show the different waves as well as the distribution of the density ρ and the pressure p. Here, we use the notation $x = x_1$ for better readability. The diaphragm is located at $x = x_{(0)}$ (adapted from Anderson, 1990, Figure 7.5).

of the contact discontinuity, which is induced by the moving shock wave propagating through a gas at rest, is governed by $p_{\mathrm{post}}/p_{\mathrm{pre}}$ as well as by a_{pre}. For an infinite pressure ratio $p_{\mathrm{post}}/p_{\mathrm{pre}} \to \infty$, the induced mass motion u_p can become supersonic, but not larger than $u_p/a_{\mathrm{post}} \to \sqrt{2/[\gamma(\gamma - 1)]} \approx 1.89$ in air.

Shock tube relations A classical and widely used example in both experiments and numerics in the context of unsteady wave motion is the so-called *shock tube*, since it contains all three types of relevant waves (shock wave, expansion wave, and contact discontinuity). We consider a closed tube which is divided in the middle into two parts by a diaphragm. A high pressure region and a low-pressure region are created by sucking out the air on one side of the diaphragm. The diaphragm breaks as soon as the pressure ratio across the diaphragm exceeds a critical limit. Figure 2.4 shows the situation after the diaphragm is broken. As a result, a moving shock wave is created which runs into the so-called *driven* section and an expansion wave is created which runs into the so-called *driver* section. Due to its motion, the shock wave creates a third feature, a contact continuity, which also propagates into the driven section. The density and pressure change continuously across the expansion fan, but are discontinuous across the shock wave. Across the contact discontinuity, only the density is discontinuous whereas the pressure stays constant.

A brief overview of the solution procedure is given, since we use the Sod shock tube (Sod, 1978) as a test case to verify the CNS solver in Section 5.5.1. A detailed derivation is given by Anderson (1990, Section 7.8). In the following, the index in brackets denotes the region of the shock tube in Figure 2.4 and not the spatial dimension as in the remainder of this work. We start from the known pressure difference $p_{(4)}/p_{(1)}$ across the diaphragm:

(1) Calculate $p_{(3)}/p_{(4)}$ by using

$$p_{(3)}/p_{(4)} = \frac{p_{(1)}}{p_{(4)}} \left[1 - \frac{(\gamma - 1)(a_{(4)}/a_{(1)})(p_{(3)}/p_{(4)} - 1)}{\sqrt{2\gamma[2\gamma + (\gamma + 1)(p_{(3)}/p_{(4)} - 1)]}} \right]^{2\gamma/(\gamma-1)}, \qquad (2.44)$$

which determines the strength of the incident shock wave. To solve Equation (2.44), an iterative process has to be established.

(2) Calculate all quantities across the shock wave using the moving shock relations (2.39), (2.40), (2.42) and (2.43).

(3) Calculate the pressure ratios across the different waves with

$$\frac{p_{(2)}}{p_{(1)}} = \frac{p_{(2)}/p_{(4)}}{p_{(1)}/p_{(4)}} = \frac{p_{(3)}/p_{(4)}}{p_{(1)}/p_{(4)}}, \qquad (2.45)$$

which determines the strength of the incident expansion wave.

(4) Calculate the thermodynamic quantities across the expansion wave by the isentropic relations

$$\frac{p_{(2)}}{p_{(1)}} = \left(\frac{\rho_{(2)}}{\rho_{(1)}}\right) = \left(\frac{T_{(2)}}{T_{(1)}}\right)^{\gamma/(\gamma-1)}. \qquad (2.46)$$

(5) Calculate the quantities inside the expansion wave with

$$\frac{a}{a_{(1)}} = 1 - \frac{\gamma - 1}{2}\left(\frac{u}{a_{(1)}}\right), \qquad (2.47a)$$

$$\frac{T}{T_{(1)}} = 1 - \left[\frac{\gamma - 1}{2}\left(\frac{u}{a_{(1)}}\right)\right]^2, \qquad (2.47b)$$

$$\frac{p}{p_{(1)}} = 1 - \left[\frac{\gamma - 1}{2}\left(\frac{u}{a_{(1)}}\right)\right]^{2\gamma/(\gamma-1)}, \qquad (2.47c)$$

$$\frac{\rho}{\rho_1} = 1 - \left[\frac{\gamma - 1}{2}\left(\frac{u}{a_{(1)}}\right)\right]^{2/(\gamma-1)}, \qquad (2.47d)$$

$$u_1 = \frac{2}{\gamma + 1}\left(a_1 + \frac{x_1}{t}\right). \qquad (2.47e)$$

Note that $u_{(3)} = u_{(2)} = u_p$ and $p_{(3)} = p_{(2)}$ hold across the contact discontinuity. The flow quantities inside the different regions of Figure 2.4 can be calculated by steps (1) to (5). In order to obtain the solution at a desired time t, the positions of the different waves are given

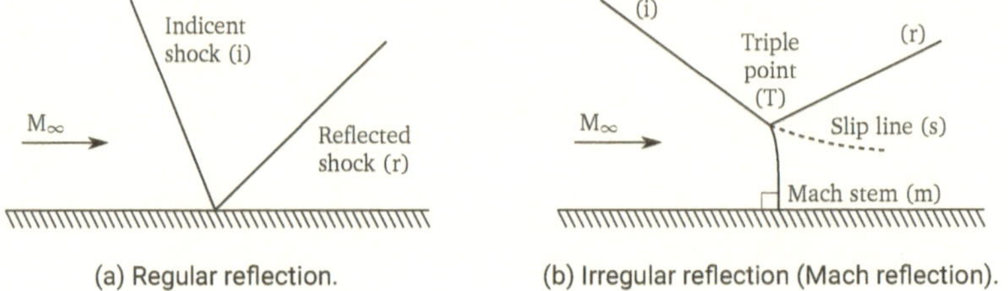

(a) Regular reflection.

(b) Irregular reflection (Mach reflection).

Figure 2.5: General types of shock wave reflections. (a) The incident shock wave hits the surface and creates a reflected shock. (b) The irregular reflection can be divided into four sub-types of which we show the standard solution of the three-shock theory (Von Neumann, 1961–1963), also known as a *Mach reflection* (adapted from Ben-Dor, 2007, Figures 1.1 and 1.2).

by

$$x_{(1)} = x_{(0)} - a_{(1)}t = x_{(0)} - \underbrace{\sqrt{\gamma \frac{p_{(1)}}{\rho_{(1)}}}}_{a_{(1)}}\, t\,, \tag{2.48a}$$

$$x_{(3)} = x_{(0)} + u_p t = x_{(0)} + \underbrace{2\frac{\sqrt{\gamma}}{\gamma - 1}\left(1 - p_{(3)}^{(\gamma-1)/(2\gamma)}\right)}_{u_p}\, t\,, \tag{2.48b}$$

$$x_{(4)} = x_{(0)} + u_s t = x_{(0)} + \underbrace{u_p \left(\frac{\rho_{(3)}/\rho_{(4)}}{\rho_{(3)}/\rho_{(4)} - 1}\right)}_{u_s}\, t\,, \tag{2.48c}$$

$$x_{(2)} = x_{(0)} + \left(u_s - a_{(2)}\right)t = x_{(0)} + \left(u_s - \underbrace{\left[a_{(1)} - \frac{\gamma - 1}{2}u_p\right]}_{a_{(2)}}\right)\, t\,, \tag{2.48d}$$

where $x_{(0)}$ denotes the position of the diaphragm (Gogol, 2020). In Equation (2.48), we have skipped the index of the spatial dimension for better readability ($x = x_1$).

2.2.4 Shock Wave Reflections

The first discovery of the shock wave reflection phenomenon is attributed to Ernst Mach (1878). He investigated two different configurations: a two-wave configuration, which is today known as a regular reflection and a three-wave configuration, which was later named after him as *Mach reflection*. We refer to Ben-Dor (2007) for a detailed introduction into the topic. In general, shock wave reflections can be divided into *regular* and *irregular reflections*, as shown in Figure 2.5. Irregular reflections are subdivided into four different types. One is the Mach reflection which is the standard solution to the three-wave theory introduced by Von Neumann (1961–1963). The regular reflection depicted in Figure 2.5a features two shock waves, the incident shock wave (i) and the reflected shock wave (r), which intersect at a reflection point

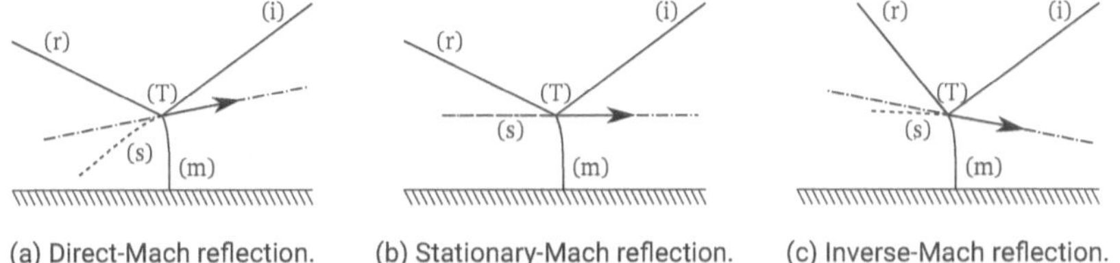

(a) Direct-Mach reflection.	(b) Stationary-Mach reflection.	(c) Inverse-Mach reflection.

Figure 2.6: Three types of Mach reflections. (a) The triple point moves away from the surface. (b) The triple point moves parallel to the surface. (c) The triple point moves towards the surface until it collides with it. After the collision, different configurations can arise (Smith, 1945; White, 1951), which are shown in Figure 2.7 (adapted from Ben-Dor, 2007, Figure 1.3).

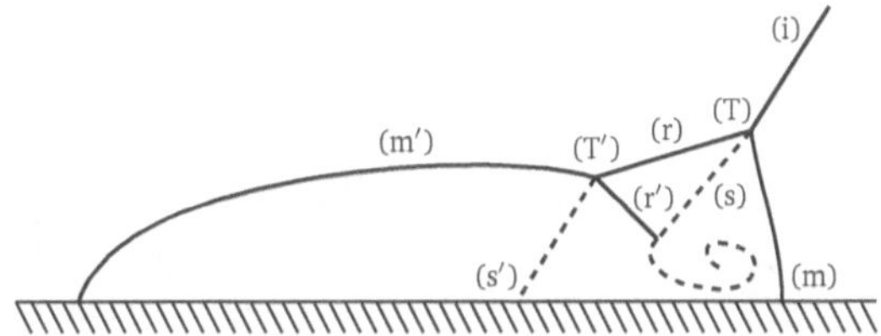

Figure 2.7: Double Mach reflection. This structure can develop if the triple point of an inverse-Mach reflection, see Figure 2.6c, hits the surface. Experimentally, it can be created by a supersonic flow over a wedge. The incident shock (i), the reflected shock (r), the Mach stem (m), and the slip line (s) meet at the first triple point (T). The secondary flow features are denoted by a prime (adapted from Kemm, 2014, Figure 1).

on the surface. The Mach reflection depicted in Figure 2.5b consists of an incident shock wave (i), a reflected shock wave (r), a Mach stem (m), and a slip line (s). All these discontinuous flow features meet at the triple point (T). The Mach stem is perpendicular to the surface at the reflection point.

The Mach reflection itself can be subdivided into three different types which depend on the propagation direction of the triple point (Courant and Friedrichs, 1948; Ben-Dor and Takayama, 1987), see Figure 2.6. We skip the Direct-Mach reflection and the Stationary-Mach reflection depicted in Figures 2.6a and 2.6b, respectively, since they are not relevant in the remainder of this work. In the case of an inverse-Mach reflection, the triple point moves towards the surface until it collides with it, see Figure 2.6c. Usually, experiments of these configurations are conducted by investigating supersonic flows over wedges or ramps. They revealed that the colliding triple points cause a deflection process of the flow at the leading edge of the wedge, for example, resulting in a complex-Mach reflection (Smith, 1945) or a *double Mach reflection (DMR)* (White, 1951). A deeper insight into the DMR, especially dealing with the influence of the initial flow field conditions and its impact on the triple point trajectories, is given by Ben-Dor (1981). In Figure 2.7, we show a schematic sketch of a DMR, which matches the test case of this work, see Section 5.5.4. The presented DMR can be created by an appropriate

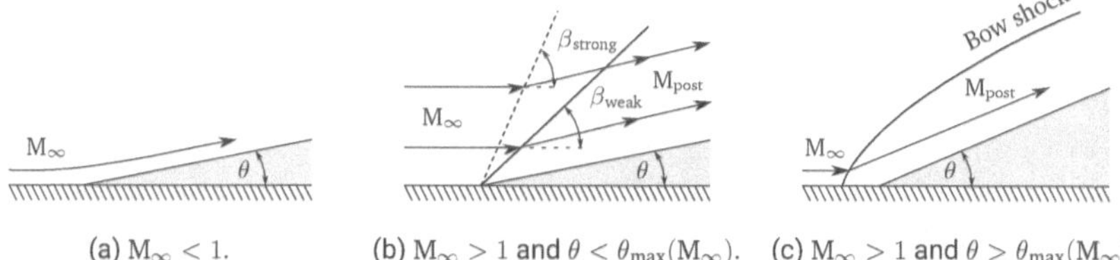

(a) $M_\infty < 1$.   (b) $M_\infty > 1$ and $\theta < \theta_{max}(M_\infty)$.   (c) $M_\infty > 1$ and $\theta > \theta_{max}(M_\infty)$.

Figure 2.8: Flow over a wedge in a steady subsonic and in a steady supersonic flow. (a) The subsonic flow ($M_\infty < 1$) turns smoothly over the wedge. (b) The supersonic flow ($M_\infty > 1$) creates an oblique shock wave which is attached to the beginning of the wedge. In general, the weak-shock solution is present where the flow behind the shock wave is supersonic. (c) The supersonic flow creates a detached bow shock in front of the wedge if the inclination angle exceeds the critical limit ($\theta > \theta_{max}(M_\infty)$) (adapted from Ben-Dor, 2007, Figure 1.6).

choice of the initial and boundary conditions. At the first triple point (T), the incident shock (i), the reflected shock (r), the Mach stem (m), and the slip line (s) meet, forming a Mach reflection. The reflected shock (r) breaks up and creates a second triple point (T′). The second reflected shock (r′) interacts with the first slip line (s) creating a curled flow structure. The second slip line (s′) hits the boundary wall and moves downstream like all other flow features, whereas solely the foot of the second Mach stem (m′) remains fixed in space.

In order to understand the physics behind shock wave reflections, we consider the simple example of a flow over a wedge for different Mach numbers M_∞ and inclination angles θ of the wedge in Figure 2.8. In Figure 2.8a, the flow is subsonic ($M_\infty < 1$) and continuously adapts itself due to the presence of the wedge. In Figure 2.8b, the flow is supersonic ($M_\infty > 1$) and the inclination angle of the wedge is below the critical angle ($\theta < \theta_{max}(M_\infty)$). In this case, an oblique shock waves forms which is attached to the beginning of the wedge. Ahead and behind the oblique shock, the flow is parallel to the surface. There are two solutions, a weak- and a strong-shock solution (Liepmann and Roshko, 2001). The shock angle of the strong-shock solution is always greater than the one for the weak-shock solution ($\beta_{strong} > \beta_{weak}$). For the strong-shock solution, the flow behind the shock is subsonic ($M_{post} < 1$), whereas it is supersonic ($M_{post} > 1$) for the weak-shock solution. It was shown in experiments that the weak-shock solution is the solution which usually occurs (Ben-Dor, 2007, Section 1.2). In Figure 2.8c, the flow is supersonic and the inclination angle of the wedge is $\theta > \theta_{max}(M_\infty)$. In this case, a detached bow shock forms in front of the wedge.

The DMR test case presented in Section 5.5.4 belongs to the category of unsteady and pseudo-steady flows. To illustrate this test case, we consider a straight incident shock wave (i) with a Mach number of M_s, which collides with a wedge inclined at an angle θ in Figure 2.9. We choose a reference frame which is fixed to the incident shock wave. The pre-shock flow on the right of the incident shock wave (i) is parallel to the wall with a Mach number of $M_\infty = M_s / \sin(\pi/2 - \theta)$. The main difference compared to the steady flow over a wedge, see in particular Figures 2.8b and 2.8c, where the post-shock flow is always supersonic due to the weak-shock solution, is that the angle of incidence $\pi/2 - \theta$ of the flow on the pre-shock side is solely controlled by the inclination angle θ of the wedge. If θ is small enough, the strong-shock solution can establish. This means that the post-shock flow can either be supersonic or subsonic ($M_{post} > 1$ or $M_{post} < 1$, respectively). If the post-shock flow is supersonic, the scenario is

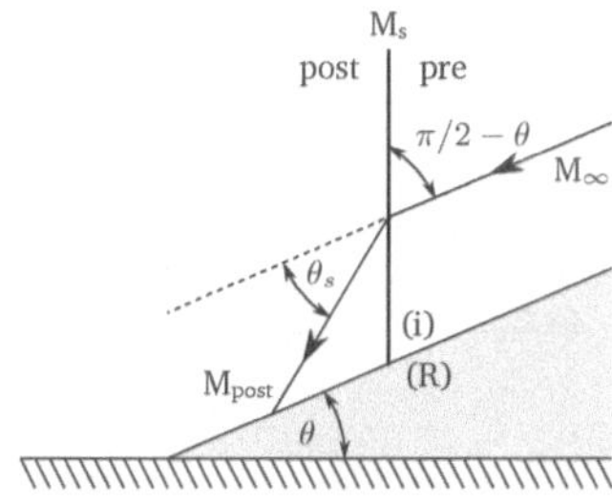

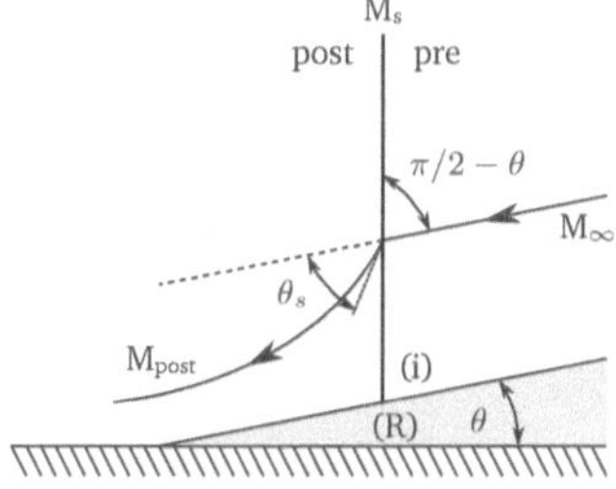

(a) Actual flow field for any value of M_{post}.

(b) Theoretically possible flow field for $M_{post} < 1$.

Figure 2.9: Shock wave reflections in unsteady and pseudo-steady flows. The deflection angle θ_s of the supersonic flow ($M_\infty > 1$) is solely governed by the inclination angle θ of the wedge. (a) If $\theta_s < \theta_{max}(M_{post})$, a regular reflection forms at the reflection point (R) and creates the depicted flow field. If $\theta_s > \theta_{max}(M_{post})$, a Mach reflection develops, resulting in the same flow field. The reflections are not drawn. (b) This configuration never occurs in reality. For any combination of M_s and θ, a Mach reflection creates the same flow field as depicted in (a) (adapted from Ben-Dor, 2007, Figure 1.8).

analogous to the steady flow over a wedge. If the deflection angle θ_s of the flow is smaller than the critical inclination angle of the wedge ($\theta_s < \theta_{max}$), an attached oblique shock wave forms emanating from the reflection point (R). We do not show this oblique shock wave in Figure 2.9a in order to avoid confusion. This newly created wave deflects the flow away from the surface and results in a regular reflection. By contrast, if $\theta_s > \theta_{max}$, a detached bow shock develops and creates a Mach reflection which deflects the flow away from the surface. Analogously to the steady flow over a wedge, see Figure 2.8a, we would expect a smoothly changing flow in Figure 2.9b if the post-shock Mach number is subsonic ($M_{post} < 1$). However, the configuration shown in Figure 2.9a arises for all combinations of M_s and θ in reality. One possible explanation is based on the fact that the flow in the theoretically possible configuration in Figure 2.9b is supersonic ahead of the incident shock (Ben-Dor, 2007, Section 1.2.2). In contrast to the steady subsonic case in Figure 2.8a, where the information of the presence of the wedge is propagated upstream, the streamline passing through the foot of the incident shock wave experiences a sudden change of the boundary condition in the unsteady case and has to adapt to it. This fact probably causes an additional shock wave, which results in a Mach reflection and establishes the flow pattern as shown in Figure 2.9a (Ben-Dor, 2007, Section 1.2.2).

3 Spatial Discretization

This chapter deals with the spatial discretization of the discontinuous Galerkin (DG) methods which are applied in this work. These methods include a generic DG method, an extended discontinuous Galerkin (XDG) method, where sharp interfaces separate the computational domain into disjoint regions, and a discontinuous Galerkin immersed boundary method (DG IBM), where the geometry is embedded into a Cartesian background grid. Interfaces and immersed boundaries are implicitly defined by the zero iso-contour of a level-set function.

The state of the art of the different DG methods is presented in Section 3.1, before we introduce the mathematical concept of a generic DG method for the application in supersonic compressible flow in Section 3.2. In Section 3.3, we extend the standard definitions for the use in an XDG method, which inherently includes the DG IBM. Thereby, we focus on important aspects, such as the choice of appropriate numerical fluxes, the implementation of boundary conditions, the quadrature in cut-cells, and a cell-agglomeration technique.

The presented DG methods are implemented in the *BoSSS* framework, see Section 1.1 for an overview of the available features. In this work, we employ the *Compressible Navier-Stokes (CNS)* solver, which relies on a standard DG method and on a DG IBM, and the *XDGShock* solver, which is based on an XDG method.

3.1 State of the Art[2]

Within the past decades, high-order DG methods have gained great popularity for the discretization of partial differential equations (PDEs) in the context of Computational Fluid Dynamics (CFD) due to their favorable properties, such as the local character, being applicable to arbitrary geometries on unstructured grids, and enabling an efficient parallelization. This makes the DG method a suitable candidate for High-Performance Computing (HPC) applications (Wang et al., 2013; Altmann et al., 2013; Dryja and Krzyżanowski, 2016).

The DG method combines many concepts from the classical Finite Volume Method (FVM) and Finite Element Method (FEM) (Hesthaven and Warburton, 2007; Di Pietro and Ern, 2012). Like the FVM the DG method is locally conservative and uses physically based flux functions, which are an essential part in the modeling of flows of any type. The DG method can be applied on arbitrary unstructured grids by using suitable basis functions, whereas in general, a high-order FVM is restricted to structured Cartesian grids. The increasing size of the stencil, representing the connection between neighboring cells, limits the parallelization and the geometrical flexibility of an FVM. The DG method can offer a high-order of accuracy like the

---

[2]Extended version of Geisenhofer et al. (2019, Section 1).

FEM, whereas it is not restricted to globally continuous basis functions. In particular, the DG method is hp-adaptive, making it possible to adapt and refine the grid locally. Additionally, a local variation of the polynomial basis functions is possible. In contrast to the FEM, having large global mass matrices which are costly to invert, the DG method features a highly local character, leading to easily invertible block-diagonal mass matrices. Especially, the combination with explicit time-integration schemes renders the DG method a highly parallelizable numerical method (Biswas et al., 1994). We refer to the work of Cockburn et al. (2000) for a detailed overview of DG methods.

Generic discontinuous Galerkin methods The origin of the DG method dates back to the early 1970's, when Reed and Hill (1973) proposed a high-order method for the solution of the neutron transport equation. Their work was supplemented by the corresponding analysis by Lesaint and Raviart (1974). They reported a convergence rate up to $O(h^P)$ for general triangular grids and $O(h^{P+1})$ for Cartesian grids with h being the characteristic grid size and P being the degree of the polynomial basis functions. Richter (1988) obtained the optimal rate $O(h^{P+1})$ for some two-dimensional structured non-Cartesian grids. Note that the convergence rate of $O(h^{P+1})$ is an optimal estimate for sufficiently smooth solutions. It is possible to construct cases where the convergence rate is limited to $O(h^{P+0.5})$ (Peterson, 1991). However, the optimal convergence rate of $O(h^{P+1})$ was observed in numerous practical applications in the context of nonlinear hyperbolic conservation laws, even though it is challenging to find an analytical expression (Cockburn et al., 2000). The reader is referred to an exemplary investigation of the DG convergence rate and its properties in Section 3.2.2, see paragraph *Spatial convergence rate*.

As the present work deals with compressible flow, we focus on the developments in the context of hyperbolic conservation laws, where the high-order Runge-Kutta discontinuous Galerkin method (RKDGM) by Cockburn and Shu (1989) can be seen as fundamental contribution. Based on this work, several extensions were published, which further attracted the attention of the DG method: Cockburn et al. (1990), Cockburn and Shu (1991), and Cockburn and Shu (1998b). Later, Cockburn and Shu (1998a) extended their RKDGM to nonlinear, time-dependent convection-diffusion systems by applying a local discontinuous Galerkin (LDG) method. Here, the convection-diffusion system is rewritten in a larger first-order system, after which it is discretized by the RKDGM.

Many works made use of approaches, which were successfully applied in the FVM and the FEM, such as space-time schemes (Van der Vegt and Van der Ven, 2002; Lörcher et al., 2007; Kummer et al., 2018), local time-stepping schemes (Hindenlang et al., 2012; Winters and Kopriva, 2014; Krämer-Eis, 2017), the construction of numerical fluxes based on Riemann solvers (Feistauer et al., 2003; Toro, 2009), and limiting procedures (Persson and Peraire, 2006; Shu, 2016).

For the solution of elliptic problems, another type of methods, so-called *(symmetric) interior penalty methods*, were mainly developed in the 1980's, first introduced by Douglas and Dupont (1976) and Arnold (1982). Further works (Shahbazi, 2005; Brezzi et al., 2006) investigated the stability in the context of DG methods, which were later applied in incompressible flows (Cockburn et al., 2005; Girault et al., 2005; Shahbazi et al., 2007).

Extended discontinuous Galerkin methods   The basic idea of extended methods on fictitious domains is an enrichment of the approximation space for PDEs with discontinuous solutions. This class of methods deals with problems where the problem domain, including the geometry representation, does not conform with the computational domain.

There are several developments, such as the Extended Finite Element Method (XFEM) (Moës et al., 1999; Fries and Belytschko, 2010), Nietzsche-type methods (Nitsche, 1971; Hansbo and Hansbo, 2002; Burman et al., 2015), the Finite Cell Method (Parvizian et al., 2007) or even the Finite Difference Method (FDM) (Brehm et al., 2015). The XFEM goes back to the work by Moës et al. (1999) in the context of crack propagation in solid mechanics. Later, adaptions to incompressible (two-phase) flows were published by Groß and Reusken (2007), Fries (2009), and Sauerland and Fries (2013).

The origin of the XDG method dates back to the work by Bastian and Engwer (2009), who focused on the discretization of two- and three-dimensional, elliptic, scalar model problems on complex-shaped domains. This work, together with other works related to high-order solutions of Poisson-type problems, share the idea of applying cell-local, piece-wise planar, triangular sub-cells for the integration of the weak forms in cut-cells. Later, Heimann et al. (2013) and Kummer (2016) applied the XDG method to incompressible flows by using a (symmetric) interior penalty discretization. Kummer (2016) proposed a combination with a sharp interface description and a cell-agglomeration technique using an enhanced quadrature scheme on cut-cells (Müller et al., 2013). Furthermore, space-time approaches for two-phase problems were developed, for example, by Lehrenfeld (2015) and Kummer et al. (2018).

To the best of our knowledge, a full XDG method for supersonic compressible flow, considering the pre-shock state and the post-shock state as two 'species', is not known until today.

Discontinuous Galerkin immersed boundary methods   Immersed boundary methods (IBMs) are closely related to the previously introduced extended methods, separating discretization and geometry representation to a large extent. In extended methods, any interface description usually splits the computational domain into a set of disjoint regions. Thus, for example, a single-phase DG IBM can be considered as an XDG method, where one region describes the fluid flow and the other region describes the embedded geometry being void.

The first IBM was originally presented by Peskin (1972) in the context of blood flows. He used a Cartesian grid whose boundary was not aligned with the geometrical topology. Later, many enhancements for several numerical schemes were published, especially for computational domains with moving or complex geometries, see Mittal and Iaccarino (2005) for an overview. However, many of these schemes are limited to second-order accuracy, at most.

The first DG IBM was published by Fidkowski and Darmofal (2007), who additionally presented a cut-cell agglomeration technique for the two-dimensional compressible Navier-Stokes equations on triangular grids. They applied a spline-based reconstruction of the embedded geometry, together with a moment-fitting-like quadrature scheme, however, with a very high number of quadrature nodes. Ongoing works were published by Modisette and Darmofal (2010) and Sun and Darmofal (2014) in the context of elliptic multi-physics problems, where no assumption other than Lipschitz continuity was made. Subsequently, Qin and Krivodonova (2013) presented a DG IBM applying an explicit time-integration scheme for smooth solutions of the Euler equations on Cartesian grids. The embedded geometry was implicitly given by the zero iso-contour of a level-set function. Especially in the case of explicit time-integration schemes, small and ill-shaped cut-cells may cause significant stability problems stemming from ill-conditioning and increase the computational load due to their severe cell-local time-step restriction. Qin and Krivodonova (2013) provided an algorithm for splitting complex cut-cells into piece-wise planar sub-cells, on which standard quadrature rules can be applied. Additionally, they applied a cell-agglomeration technique, combined with a special adiabatic slip-wall boundary condition at the curved interface. As a result, they regained a convergence rate up to third order.

Müller et al. (2017) generalized the work of Qin and Krivodonova (2013) and extended this approach to the compressible Navier-Stokes equations by employing a Hierarchical Moment-Fitting (HMF) strategy (Müller et al., 2013) for the integration in cut-cells. The HMF strategy uses an hierarchical evaluation of integrals on lower-dimensional sub-domains.

Recently, Geisenhofer et al. (2019) proposed a DG IBM for supersonic compressible flow, where discontinuities, such as shock waves, were treated with a two-step shock-capturing strategy based on artificial viscosity (Persson and Peraire, 2006; Persson, 2013). Additionally, an adaptive local time-stepping strategy was implemented (Winters and Kopriva, 2014; Krämer-Eis, 2017) in order to regain computational performance, since small cut-cells with an active second-order artificial viscosity term restrict the maximum admissible time-step size drastically.

3.2 A Generic Discontinuous Galerkin Method

A detailed introduction to a generic DG method is given, for example, in the textbooks by Cockburn (2003), Li (2006), Hesthaven and Warburton (2007), and Di Pietro and Ern (2012). For an introduction in the context of the underlying *BoSSS* framework, the reader is referred to the recent works by Geisenhofer et al. (2019), Kummer et al. (2020), Kummer and Müller (2020), Kikker (2020), and Smuda (2020). For the sake of consistency and comprehensibility, we stick to the notation of these works and follow the derivations presented therein.

3.2.1 Definitions

In this section, we introduce several basic definitions for the discretization of a D-dimensional computational domain $\Omega \subset \mathbb{R}^D$ which has to be polygonal and simply connected (Kummer et al., 2020; Smuda, 2020). In this work, we restrict ourselves to one- and two-dimensional problems $D \in \{1, 2\}$.

Definition 3.1 (Basic notations). We introduce the following definitions:

- The computational domain Ω is divided into a set of cells $\mathcal{K}_h = \{K_1, \ldots, K_J\}$, where h denotes the characteristic length scale of the numerical grid. The cells are non-overlapping $\int_{K_j \cap K_k} 1 \, dV = 0$ for all $j \neq k$ and cover the entire computational domain $\overline{\Omega} = \cup_j \overline{K_j}$.

- On the computational domain Ω, we define a set of all inner and outer edges $\Gamma = \Gamma_{\text{int}} \cup \Gamma_{\text{out}} = \cup_j \partial K_j$, where ∂K_j denotes the boundary of a grid cell. We write the set of inner edges as $\Gamma_{\text{int}} = \Gamma \setminus \partial \Omega$, where $\partial \Omega$ refers to the boundary of the computational domain. The boundary edges can be subdivided into edges with Dirichlet and Neumann boundary conditions $\Gamma_{\text{out}} = \Gamma_{\text{Dir}} \cup \Gamma_{\text{Neu}}$.

- On all edges Γ, we introduce a field of normal vectors $\mathbf{n}_\Gamma$. On $\partial\Omega$, it represents the outwards pointing normals $\mathbf{n}_\Gamma = \mathbf{n}_{\partial\Omega}$.

- We denote the broken gradient for a smooth scalar field $\psi \in C^1(\Omega \setminus \Gamma)$ as $\nabla_h \psi$. In the same manner, we denote the broken divergence as $\nabla_h \cdot \psi$.

- We define the inner product for two vectors $\mathbf{a}, \mathbf{b} \in \Omega \subset \mathbb{R}^D$ in D dimensions as $\langle \mathbf{a}, \mathbf{b} \rangle := \mathbf{a} \cdot \mathbf{b} := \sum_{d=1}^{D} a_d b_d$. This induces the ℓ^2-norm $|\mathbf{a}|_2 := \sqrt{\langle \mathbf{a}, \mathbf{a} \rangle} = \left(\sum_{d=1}^{D} a_d^2 \right)^{1/2}$.

- Similarly, we define the inner product for two functions $\langle \psi, \vartheta \rangle := \int_{\mathbf{x} \in \Omega} \psi(\mathbf{x}) \vartheta(\mathbf{x}) \, dV$ on $\Omega \subset \mathbb{R}^D$ in D dimensions. This induces the L^2-norm $\|\psi\|_2 := \|\psi\|_{L^2(\Omega \subset \mathbb{R}^D)} := \sqrt{\langle \psi, \psi \rangle} = \left(\int_{\mathbf{x} \in \Omega} (\psi(\mathbf{x}))^2 \, dV \right)^{1/2}$. In addition, two functions are orthogonal, if and only if $\langle \psi, \vartheta \rangle = 0$.

Definition 3.2 (Discontinuous Galerkin (DG) space). We define the broken polynomial DG space as

$$\mathbb{P}_P(\mathcal{K}_h) := \left\{ f \in L^2(\Omega); \forall K \in \mathcal{K}_h : f|_K \text{ is polynomial and } \deg(f|_K) \leq P \right\} \tag{3.1}$$

with a total polynomial degree of P.

Definition 3.3 (Inner and outer values, average and jump operator). For a scalar field $\psi \in C^0(\Omega \setminus \Gamma)$, we write the inner and outer values at edges Γ as

$$\psi^{\text{in}}(\mathbf{x}) := \lim_{\xi \searrow 0} \psi(\mathbf{x} - \xi \mathbf{n}_\Gamma), \qquad \mathbf{x} \in \Gamma, \tag{3.2a}$$

$$\psi^{\text{out}}(\mathbf{x}) := \lim_{\xi \searrow 0} \psi(\mathbf{x} + \xi \mathbf{n}_\Gamma), \qquad \mathbf{x} \in \Gamma_{\text{int}}. \tag{3.2b}$$

We introduce the jump operator

$$[\![\psi]\!] := \begin{cases} \psi^{\text{in}} - \psi^{\text{out}}, & \mathbf{x} \in \Gamma_{\text{int}}, \\ \psi^{\text{in}}, & \mathbf{x} \in \Gamma_{\text{out}}, \end{cases} \tag{3.3}$$

and the average operator

$$\{\!\{\psi\}\!\} := \begin{cases} \left(\psi^{\text{in}} + \psi^{\text{out}} \right) / 2, & \mathbf{x} \in \Gamma_{\text{int}}, \\ \psi^{\text{in}}, & \mathbf{x} \in \Gamma_{\text{out}}. \end{cases} \tag{3.4}$$

3.2.2 Derivation

In this section, we derive the DG discretization of a generic scalar conservation law with a nonlinear flux function (Hesthaven and Warburton, 2007, Section 2.2; Smuda, 2020).

Test problem   We consider a scalar field $\psi = \psi(\mathbf{x}, t)$ on the computational domain Ω with suitable initial conditions ψ_0 and suitable Dirichlet boundary conditions $\Gamma_{\text{out}} = \Gamma_{\text{Dir}}$. We state a generic conservation law for $\psi(\mathbf{x}, t)$ with a nonlinear flux function $\mathbf{f}(\psi)$ as

$$\frac{\partial \psi}{\partial t} + \nabla \cdot \mathbf{f}(\psi) = 0\,, \qquad\qquad \mathbf{x} \in \Omega\,, \tag{3.5a}$$

$$\psi = \psi_{\text{Dir}}\,, \qquad\qquad \mathbf{x} \in \Gamma_{\text{Dir}}\,, \tag{3.5b}$$

$$\psi(\mathbf{x}, t_0) = \psi_0(\mathbf{x})\,, \qquad\qquad \mathbf{x} \in \Omega\,. \tag{3.5c}$$

Cell-local approximation   Equation (3.5) is defined in a continuous setting. We look for a numerical candidate $\psi_h(\mathbf{x}, t) \approx \psi(\mathbf{x}, t) \in \Omega$ on a discrete set of cells $\mathcal{K}_h$. In each cell $K_j \in \mathcal{K}_h$, we introduce a cell-local polynomial basis vector

$$\boldsymbol{\phi}_j = (\phi_{j,n})_{n=1,\dots,N} \in \mathbb{P}_P\left(\{K_j\}\right)\,, \tag{3.6}$$

where j and n are the running indices of the cells and of the number of the polynomial basis functions, respectively. In particular, N denotes the total number of terms in the polynomial expansion and J will denote the total number of cells later on. The basis (3.6) is only non-zero in its associated cell ($\operatorname{supp}(\boldsymbol{\phi}_j) = \overline{K}_j$). We approximate the cell-local solution by

$$\psi_j(\mathbf{x}, t) = \sum_{n=1}^{N} \tilde{\psi}_{j,n}(t)\, \phi_{j,n}(\mathbf{x}) = \tilde{\boldsymbol{\psi}}_j(t) \cdot \boldsymbol{\phi}_j(\mathbf{x})\,, \qquad \mathbf{x} \in K_j\,, \tag{3.7}$$

where

$$\tilde{\boldsymbol{\psi}}_j = (\tilde{\psi}_{j,n})_{n=1,\dots,N} \tag{3.8}$$

denotes the time-dependent, unknown degrees of freedom (DOF). In the *BoSSS* framework, a modal orthogonal basis consisting of Legendre polynomials is used, which satisfies the orthogonality condition

$$\int_{K_j} \phi_{j,m} \phi_{j,n}\, \mathrm{d}V = \delta_{mn}\,, \tag{3.9}$$

where δ_{mn} denotes the Kronecker delta.

Cell-local residual   We insert Equation (3.7) into Equation (3.5a) in order to obtain a discrete cell-local approximation of the test problem. This leads to the cell-local residual

$$R_j(\mathbf{x}, t) = \frac{\partial \psi_j}{\partial t} + \nabla \cdot \mathbf{f}(\psi_j)\,, \qquad \mathbf{x} \in K_j\,. \tag{3.10}$$

We minimize the residual with respect to the space of trial functions $\mathbb{P}_P\left(\{K_j\}\right)$. Following the Galerkin procedure, we use identical trial and test functions ($\phi_{j,m} = \vartheta_{j,m}$), multiply Equation (3.10) by $\vartheta_{j,m}$ and integrate over one cell K_j, which yields

$$\int_{K_j} R_j\, \vartheta_{j,m}\, \mathrm{d}V = \int_{K_j} \frac{\partial \psi_j}{\partial t}\, \phi_{j,m} + \nabla \cdot \mathbf{f}(\psi_j)\, \phi_{j,m}\, \mathrm{d}V \overset{!}{=} 0\,. \tag{3.11}$$

From a geometrical perspective, we enforce the residual to be orthogonal to $\mathbb{P}_P\left(\{K_j\}\right)$. Note that Equation (3.11) already defines a cell-local, linear equation system with N unknowns.

Global approximation In order to obtain a global approximation $\psi_h(\mathbf{x}, t) \approx \psi(\mathbf{x}, t) \in \Omega$, we have to incorporate all cell-local formulations (3.11) into a global one. Thus, we define

$$\psi_h(\mathbf{x}, t) = \sum_{j=1}^{J} \psi_j(\mathbf{x}, t) = \sum_{j=1}^{J} \sum_{n=1}^{N} \tilde{\psi}_{j,n}(t)\, \phi_{j,n}(\mathbf{x}) \ \in \mathbb{P}_P\left(\mathcal{K}_h\right), \tag{3.12}$$

as the superposition of all cell-local approximations (3.7), consisting of the sum of J piece-wise polynomial solutions $\psi_j(\mathbf{x}, t)$.

Semi-discrete (weak) formulation Following the typical derivation of a DG method, we perform a spatial integration by parts of Equation (3.11), which results in

$$\int_{K_j} \frac{\partial \psi_j}{\partial t} \phi_{j,m}\, \mathrm{d}V - \int_{K_j} \mathbf{f}(\psi_j) \cdot \nabla_h \phi_{j,m}\, \mathrm{d}V + \oint_{\partial K_j} (\mathbf{f}(\psi_j) \cdot \mathbf{n}_\Gamma)\, \phi_{j,m}\, \mathrm{d}S = 0. \tag{3.13}$$

The surface integral in Equation (3.13) is multiply defined at the internal cell boundaries Γ_{int} due to the definition of the basis (3.6). To obtain a unique solution on Γ_{int}, we introduce a numerical flux

$$\hat{F}\left(\psi_h^{\text{in}}, \psi_h^{\text{out}}, \mathbf{n}_\Gamma\right) \approx \mathbf{f}(\psi_j) \cdot \mathbf{n}_\Gamma, \tag{3.14}$$

which combines the solution on both sides in a suitable way. Details about the numerical fluxes in the context of the Euler equations (2.5) can be found in Section 3.2.3. Based on Equation (3.13), we formulate the global semi-discrete form on $\mathcal{K}_h$ as a global minimization problem for $\psi_h(\mathbf{x}, t)$, $\mathbf{x} \in \Omega$ as follows: Find $\psi_h(\mathbf{x}, t) \in \mathbb{P}_P(\mathcal{K}_h)$ such that

$$\int_{\Omega} \frac{\partial \psi_h}{\partial t} \phi\, \mathrm{d}V - \int_{\Omega} \mathbf{f}(\psi) \cdot \nabla_h \phi\, \mathrm{d}V + \oint_{\Gamma} \hat{F}\left(\psi_h^{\text{in}}, \psi_h^{\text{out}}, \mathbf{n}_\Gamma\right) [\![\phi]\!]\, \mathrm{d}S = 0. \tag{3.15}$$

Equation (3.15) has been discretized in space, but not in time. Details about the time discretization procedure can be found in Chapter 4. It is common to write Equation (3.15) in a shorthand form as

$$\mathbf{M}\frac{\mathrm{d}\tilde{\psi}}{\mathrm{d}t} + \mathbf{Op}(\tilde{\psi}) = \mathbf{b}, \tag{3.16}$$

where $\mathbf{M} \in \mathbb{R}^{N,N}$ denotes the symmetric and block-diagonal mass matrix

$$\mathbf{M} := \begin{bmatrix} \mathbf{M}_1 & 0 & \dots & 0 \\ 0 & \mathbf{M}_2 & \dots & 0 \\ \vdots & \vdots & \ddots & \vdots \\ 0 & 0 & \dots & \mathbf{M}_J \end{bmatrix}. \tag{3.17}$$

In Equation (3.17) each block corresponds to a cell-local mass matrix

$$(\mathbf{M}_j)_{m,n} := \int_{K_j} \phi_{j,m}\phi_{j,n}\, \mathrm{d}V. \tag{3.18}$$

The global vector of unknowns is given by

$$\tilde{\psi} := \left\{\tilde{\psi}_{1,1}, \tilde{\psi}_{1,2}, \dots, \tilde{\psi}_{j,n}, \dots, \tilde{\psi}_{J,N}\right\} \in \mathbb{R}^N. \tag{3.19}$$

	$h = 1/2$	$h = 1/8$	$h = 1/32$	$h = 1/64$
$P = 1$	Ref. $\mid$ 1.0 s	-	$\mathbf{5.7 \cdot 10^{-3} \mid 19.6\,s}$	$1.4 \cdot 10^{-3} \mid 54.3\,s$
$P = 2$	-	$\mathbf{6.3 \cdot 10^{-3} \mid 7.3\,s}$	-	$1.3 \cdot 10^{-5} \mid 110.0\,s$
$P = 4$	$\mathbf{3.3 \cdot 10^{-3} \mid 4.9\,s}$	-	-	$3.3 \cdot 10^{-10} \mid 327.0\,s$
$P = 8$	$2.1 \cdot 10^{-7} \mid 57.8\,s$	$4.8 \cdot 10^{-12} \mid 279.0\,s$	$5.0 \cdot 10^{-13} \mid 1958.0\,s$	$6.6 \cdot 10^{-13} \mid 5256.0\,s$

Table 3.1: Global L^2-errors $\|\psi - \psi_h\|_2$ and execution times in s for several polynomial degrees P and grid sizes h for the test problem (3.21). Approximately, the same moderate L^2-error can be obtained for different combinations of P and h (marked in bold). The high-order solution is the best in terms of the execution time (shaded in gray). Very low L^2-errors can only be obtained for large P in a reasonable setting (last row) (adapted from Hesthaven and Warburton, 2007, Tables 2.1 to 2.3).

Based on Equations (3.13) and (3.14), we define the cell-local discrete operator as

$$(\mathbf{Op}_j)_m := - \int_{K_j} \mathbf{f}(\psi_j) \cdot \nabla_h \phi_{j,m} \, \mathrm{d}V + \oint_{\partial K_j} \hat{F}\left(\psi_h^{\mathrm{in}}, \psi_h^{\mathrm{out}}, \mathbf{n}_\Gamma\right) \phi_{j,m} \, \mathrm{d}S. \tag{3.20}$$

In Equation (3.16), the right hand side vector $\mathbf{b}$ contains the Dirichlet boundary conditions (3.5b).

Spatial convergence rate   We consider the one-dimensional test problem

$$\frac{\partial \psi(x_1, t)}{\partial t} - 2\pi \frac{\partial \psi(x_1, t)}{\partial x_1} = 0, \qquad x_1 \in [0, 2\pi], \tag{3.21}$$

using periodic boundary conditions. The initial conditions are set to $\psi(x_1, 0) = \sin(l x_1)$ with $l = 2\pi/\hat{\lambda}$, where $\hat{\lambda}$ is the wave length, following Hesthaven and Warburton (2007, Section 2.2). There are two ways to investigate the spatial convergence rate, while neglecting temporal errors. Hence, Hesthaven and Warburton (2007) set a very small and constant time-step size for all cases. First, the spatial resolution h can be increased. Second, the degree P of the polynomial approximation can be increased. The spatial convergence rate is given by

$$\|\psi - \psi_h\|_2 \leq C h^{P+1}, \tag{3.22}$$

where C denotes a constant which does not depend on h. Note that increasing P gives a faster convergence rate than increasing h. Equation (3.22) only holds for sufficiently smooth functions $\psi(\mathbf{x}, t)$. The execution time and, thus, the computational effort scale like

$$\text{Execution time} \simeq C(t_{\mathrm{end}}) h (P + 1)^2. \tag{3.23}$$

The constant $C(t_{\mathrm{end}})$ depends on the end time and the time-step size. At this point, Hesthaven and Warburton (2007) neglect that the time-step size also depends on h and P. Equation (3.23) makes high-order solutions not a suitable candidate because of the large execution times, see also Table 3.1. However, for a moderate accuracy ($\|\psi - \psi_h\|_2 \approx 5.0 \cdot 10^{-3}$), the high-order solution outperforms comparable low-order runs in terms of the execution time. Without going into the details, however, Hesthaven and Warburton (2007) show that large values of P are beneficial for long-time computations.

Properties of the discontinuous Galerkin method   If the numerical flux (3.14) fulfills several requirements, it can be shown that the DG method, written in the form of Equation (3.15), is consistent and locally conservative by construction (Cockburn, 2003). The choice of the flux function has a strong impact on the stability and accuracy of the DG method and depends heavily on the discretized PDE, see Section 3.2.3 for details.

Summing up, we state the main properties of the DG method (Hesthaven and Warburton, 2007, Section 2.2):

- The numerical solutions are piece-wise smooth, but they are discontinuous between neighboring cells.

- The boundary conditions and the continuity at cell boundaries are only enforced in a weak sense.

- All operators are local.

- The DG method is well suited for local hp-adaptivity, since the information exchange is only between neighboring cells. Additionally, this enables an efficient parallelization.

- The latter point makes the DG method a suitable candidate for HPC applications.

3.2.3 Numerical Fluxes

After the introduction to a generic DG method in Sections 3.2.1 and 3.2.2, we present essential properties of the numerical fluxes, which are elementary parts for any stable and converging DG method (Di Pietro and Ern, 2012, Section 3.2). A detailed discussion of suitable numerical fluxes for hyperbolic conservation laws can be found in LeVeque (1999) and Toro (2009). Subsequently, we briefly summarize the convective and diffusive numerical fluxes applied in this work, following Kummer and Müller (2020) and Smuda (2020). We repeat the notation of the numerical fluxes for the sake of convenience.

$$\hat{F}\left(\psi_h^{\mathrm{in}}, \psi_h^{\mathrm{out}}, \mathbf{n}_\Gamma\right) \approx \mathbf{f}(\psi_j) \cdot \mathbf{n}_\Gamma . \tag{3.14, repeated}$$

Stability   We define the stability via the energy estimate in a continuous setting on the computational domain Ω as

$$\|\psi(\mathbf{x}, t)\|_2 \leq \|\psi(\mathbf{x}, 0)\|_2 , \qquad \mathbf{x} \in \Omega , \; t \geq 0 , \tag{3.24}$$

assuming homogeneous Dirichlet boundary conditions. A DG method is said to be stable if the norm $\|\psi(\mathbf{x}, t)\|_2$ does not increase in time.

Lipschitz continuity   This assumption on the numerical flux is necessary in order to prove the existence and the uniqueness of the semi-discrete form (3.15). The sufficient condition is that the numerical flux function $\hat{F}(a, b, \mathbf{n})$ is differentiable in the first and second argument and the magnitudes of the derivatives $\partial_a \hat{F}(a, b, \mathbf{n})$ and $\partial_b \hat{F}(a, b, \mathbf{n})$ are bounded, yielding the expressions

$$\exists C_a \in \mathbb{R} : |\hat{F}(a_1, b, \mathbf{n}) - \hat{F}(a_2, b, \mathbf{n})| \leq C_a |a_1 - a_2| , \qquad a_1, a_2 \in \mathbb{R} , \tag{3.25a}$$

$$\exists C_b \in \mathbb{R} : |\hat{F}(a, b_1, \mathbf{n}) - \hat{F}(a, b_2, \mathbf{n})| \leq C_b |b_1 - b_2| , \qquad b_1, b_2 \in \mathbb{R} . \tag{3.25b}$$

Consistency   Another requirement on the numerical flux is to meet the consistency property

$$\hat{F}(a, a, \mathbf{n}) = \mathbf{f}(a) \cdot \mathbf{n}, \qquad a \in \mathbb{R}, \tag{3.26}$$

which additionally requires that $\hat{F}$ must be single-valued across cell boundaries. The proof uses the strong form, which is obtained by partially integrating the weak form (3.15) once more. If the numerical flux is consistent, it directly follows that the weak form is fulfilled when inserting the continuous solution ($\psi_h = \psi$).

Monotonicity   The semi-discrete form (3.15) fulfills the discrete equivalent of the energy estimate (3.24) if the numerical flux is Lipschitz continuous and monotone

$$\frac{\partial \hat{F}(a, b, \mathbf{n})}{\partial a} \geq 0 \quad \wedge \quad \frac{\partial \hat{F}(a, b, \mathbf{n})}{\partial b} \leq 0, \qquad a, b \in \mathbb{R}, \tag{3.27}$$

which is a necessary condition for stability.

Conservativity   The DG method has to meet this physical requirement in a global sense. The total amount of the conserved scalar quantity $\psi(\mathbf{x}, t)$ on the domain Ω changes only due to the fluxes across the domain boundary $\partial\Omega$. Thus, the numerical flux has to fulfill

$$\hat{F}(a, b, \mathbf{n}) = -\hat{F}(a, b, -\mathbf{n}), \qquad a, b \in \mathbb{R}. \tag{3.28}$$

Riemann problem   The construction and evaluation of numerical fluxes (3.14) are closely related to the approximate solution of the simplest, non-trivial *Riemann problem* (Riemann, 1860) which is given by

$$\frac{\partial \mathbf{U}}{\partial t} + \frac{\partial \mathbf{F}_1(\mathbf{U})}{\partial x_1} = 0, \tag{3.29}$$

with the discontinuous initial conditions

$$\mathbf{U}(x_1, 0) = \begin{cases} \mathbf{U}_\mathrm{L}, & \text{if } x_1 < 0, \\ \mathbf{U}_\mathrm{R}, & \text{if } x_1 > 0. \end{cases} \tag{3.30}$$

where $(\cdot)_\mathrm{L}$ and $(\cdot)_\mathrm{R}$ denote the left- and right-sided quantities. Note that Equations (3.29) and (3.30) correspond to a one-dimensional initial-value problem. The domain of interest is usually visualized in the x_1-t-plane and consists of points (x_1, t) with $-\infty < x_1 < \infty$ and $t > 0$.

In the following, we briefly introduce the relevant theory of the Riemann problem mainly based on Toro (2009, Chapters 3, 4, and 9). The Riemann problem is useful due to the following properties:

- It is the solution to a system of hyperbolic conservation laws for the simplest, non-trivial initial conditions (3.30).

- The solution of the Riemann problem contains all relevant physical and mathematical properties of the underlying conservation laws.

- In case of the Euler equations, the Riemann problem is equivalent to the shock tube problem in the limit of an infinitely long tube, see Section 2.2.3.

- The Riemann problem is widely used as a reference for numerical methods. Even though there is no exact solution in closed form for the Euler equations, the Riemann problem can be iteratively solved up to any desired accuracy.

- The global solution can be formed as the superposition of local Riemann problems across cell boundaries. Godunov (1959) proposed the first Riemann solver for the Euler equations, which can be seen as a first-order FVM variant from today's point of view. There are two ways of extracting information from the Riemann problem which can be used in Godunov-type methods: First, we look for an *approximation to the numerical flux* and second, we look for an *approximation to a state* and then evaluate the physical flux at this state (Toro, 2009, Chapter 9).

- Examples for approximate Riemann solvers were published by Engquist and Osher (1980) and Roe (1981). The approximate solution of the Riemann problem is favorable due to the following arguments (Salas, 2010, Section 2.3.13):

  - The exact Riemann solver provides more information than necessary in the case of the Euler equations.

  - The extrapolation of the solution at the cell boundaries is approximated. Thus, an exact solution does not increase the accuracy due to inexact initial conditions.

  - The exact Riemann solver is computationally expensive for the Euler equations.

  - There is no Riemann solver for the multi-dimensional problem.

In order to link the Riemann problem to the Euler equations and to show several useful properties, we first write the one-dimensional Euler equations in their quasi-linear form as

$$\frac{\partial \mathbf{U}}{\partial t} + \mathbf{F}'(\mathbf{U})\frac{\partial \mathbf{U}}{\partial x_1} = 0 \,, \tag{3.31}$$

where $\mathbf{F}'(\mathbf{U})$ is the Jacobian matrix

$$\mathbf{F}'(\mathbf{U}) = \frac{\partial F_i}{\partial U_j} = \begin{bmatrix} 0 & 1 & 0 \\ -\frac{1}{2}(\gamma-3)\left(\frac{U_2}{U_1}\right)^2 & (3-\gamma)\left(\frac{U_2}{U_1}\right) & \gamma-1 \\ -\frac{\gamma U_2 U_3}{U_1^2} + (\gamma-1)\left(\frac{U_2}{U_1}\right)^2 & \frac{\gamma U_3}{U_1} - \frac{3}{2}(\gamma-1)(\frac{U_2}{U_1})^2 & \gamma\left(\frac{U_2}{U_1}\right) \end{bmatrix} \tag{3.32}$$

with the vector of conserved quantities $\mathbf{U} = (U_1, U_2, U_3)^\mathsf{T} = (\rho, \rho u_1, E)^\mathsf{T}$. The eigenvalues of $\mathbf{F}'(\mathbf{U})$ are given by

$$\lambda_1 = u_1 - a \,, \qquad \lambda_2 = u_1 \,, \qquad \lambda_3 = u_1 + a \,, \tag{3.33}$$

where u_1 denotes the velocity of the fluid and a denotes the speed of sound. The corresponding right eigenvectors ($\mathbf{F}'\mathbf{v} = \lambda\mathbf{v}$) are given by

$$\mathbf{v}^{(1)} = \begin{bmatrix} 1 \\ u_1 - a \\ (E+p)/\rho - u_1 a \end{bmatrix}, \qquad \mathbf{v}^{(2)} = \begin{bmatrix} 1 \\ u_1 \\ \frac{1}{2}u_1^2 \end{bmatrix}, \qquad \mathbf{v}^{(3)} = \begin{bmatrix} 1 \\ u_1 + a \\ (E+p)/\rho + u_1 a \end{bmatrix} . \tag{3.34}$$

All eigenvalues (3.33) are real, and the corresponding eigenvectors (3.34) form a complete set of linearly independent eigenvectors. This proves that the time-dependent, one-dimensional Euler equations in combination with the ideal gas law (2.9) are hyperbolic (Toro, 2009, Section 3.1.1). The proof can be extended to higher dimensions (Toro, 2009, Section 3.2.1).

For numerical analysis, it can be helpful to use formulations in primitive variables. We rewrite the quasi-linear form (3.31) using the state vector of primitive variables $\mathbf{W} = (\rho, u_1, p)^\mathsf{T}$ as

$$\frac{\partial \mathbf{W}}{\partial t} + \mathbf{F}'(\mathbf{W})\frac{\partial \mathbf{W}}{\partial x_1} = 0, \tag{3.35}$$

with the Jacobian matrix

$$\mathbf{F}'(\mathbf{W}) = \begin{bmatrix} u_1 & \rho & 0 \\ 0 & u_1 & 1/\rho \\ 0 & \rho a^2 & u_1 \end{bmatrix}. \tag{3.36}$$

The corresponding eigenvalues are given by $\lambda_1 = u_1 - a$, $\lambda_2 = u_1$, and $\lambda_3 = u_1 + a$ as well as the corresponding left eigenvectors by

$$\mathbf{v}_\mathrm{L}^{(1)} = \begin{bmatrix} 0 \\ 1 \\ -1/(\rho a) \end{bmatrix}, \qquad \mathbf{v}_\mathrm{L}^{(2)} = \begin{bmatrix} 1 \\ 0 \\ -1/a^2 \end{bmatrix}, \qquad \mathbf{v}_\mathrm{L}^{(3)} = \begin{bmatrix} 0 \\ 1 \\ 1/(\rho a) \end{bmatrix}. \tag{3.37}$$

A characteristic equation states that along a characteristic line $\mathrm{d}x_1/\mathrm{d}t = \lambda_i$ the following property

$$\mathbf{v}_\mathrm{L}^{(i)} \cdot \begin{bmatrix} \mathrm{d}\rho \\ \mathrm{d}u_1 \\ \mathrm{d}p \end{bmatrix} = 0 \tag{3.38}$$

holds, which we directly write for the PDEs (3.35). Expanding Equation (3.38) yields the three characteristic equations of the quasi-linear form in primitive variables as

$$\begin{aligned} \mathrm{d}p - \rho a\,\mathrm{d}u_1 = 0 \quad & \text{along} \quad \mathrm{d}x_1/\mathrm{d}t = \lambda_1 = u_1 - a\,, & \text{(3.39a)} \\ \mathrm{d}p - a^2\,\mathrm{d}\rho = 0 \quad & \text{along} \quad \mathrm{d}x_1/\mathrm{d}t = \lambda_2 = u_1\,, & \text{(3.39b)} \\ \mathrm{d}p + \rho a\,\mathrm{d}u_1 = 0 \quad & \text{along} \quad \mathrm{d}x_1/\mathrm{d}t = \lambda_3 = u_1 + a\,. & \text{(3.39c)} \end{aligned}$$

Figure 3.1 shows the solution of the Riemann problem as a set of elementary waves, which are expansion waves, contact discontinuities, and shock waves. The three waves are related to three characteristic fields with the eigenvalues $u_1 - a$, u_1, and $u_1 + a$ as well as the corresponding eigenvectors $\mathbf{v}_\mathrm{L}^{(i)}$, $i = 1, 2, 3$. The waves separate four distinct states $\mathbf{U}_\mathrm{L}$, $\mathbf{U}_{*\mathrm{L}}$, $\mathbf{U}_{*\mathrm{R}}$, and $\mathbf{U}_\mathrm{R}$. The type of the left and right wave depends on the initial conditions (3.30), whereas the middle wave is always a contact discontinuity. Figure 3.2 shows the four physically relevant scenarios.

Godunov's scheme The original Godunov's scheme dates back to the 1950's, where Godunov (1959) proposed a conservative, first-order scheme in space and time for the solution of nonlinear hyperbolic conservation laws. The core is the solution of the one-dimensional Riemann problem. Godunov's scheme can also be seen as a conservative, first-order FVM, since it assumes a piece-wise constant distribution of the data. Consequently, a local Riemann

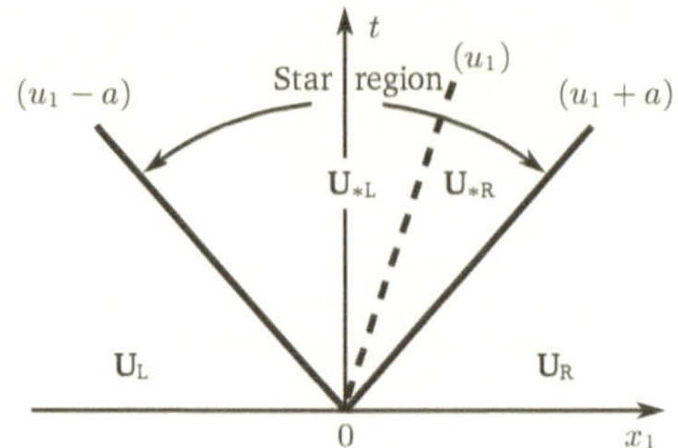

Figure 3.1: Structure of the solution $\mathbf{U}(x_1, t)$ of the Riemann problem for the one-dimensional Euler equations. Three waves with the eigenvalues $u_1 - a$, u_1, and $u_1 + a$ separate the four states $\mathbf{U}_L$, $\mathbf{U}_{*L}$, $\mathbf{U}_{*R}$, and $\mathbf{U}_R$. The type of the left and the right wave depends on the physical conditions (3.30), whereas the type of the middle wave is always a contact discontinuity (adapted from Toro, 2009, Figure 3.1).

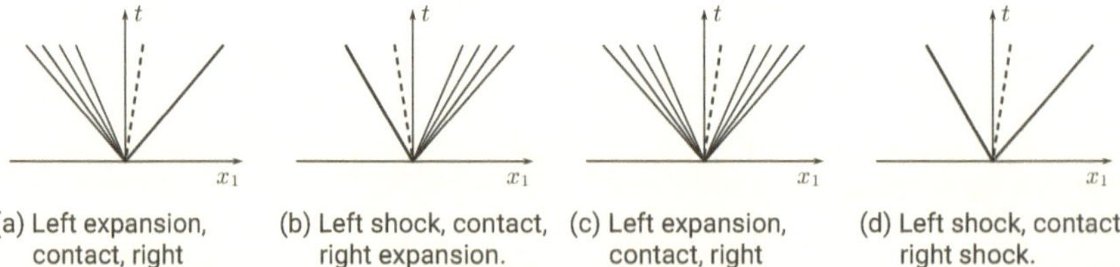

(a) Left expansion, contact, right shock.

(b) Left shock, contact, right expansion.

(c) Left expansion, contact, right expansion.

(d) Left shock, contact, right shock.

Figure 3.2: Physically relevant wave patterns of the solution $\mathbf{U}(x_1, t)$ of the Riemann problem for the one-dimensional Euler equations, see also Figure 3.1 (adapted from Toro, 2009, Figure 4.2).

problem has to be solved at every cell boundary. The solution is a similarity solution in the context of the Euler equations and depends on the ratio x_1/t. The wave speeds determine the time-step size so that the waves do not interact. Then, the global solution is obtained by a superposition of the local solutions (Toro, 2009, Chapter 6). We state the final Godunov flux in the previously introduced notation, see also Equation (3.14), as

$$\hat{F}\left(\psi_h^{\text{in}}, \psi_h^{\text{out}}, \mathbf{n}_\Gamma\right) = \mathbf{f}\left(\text{RS}(\psi_h^{\text{in}}, \psi_h^{\text{out}})\right) \cdot \mathbf{n}_\Gamma \,. \tag{3.40}$$

It reflects the 'best' approximation to the continuous flux function $\mathbf{f}(\psi_h)$ which always ensures stability. The Godunov flux is stable, physical, and hardly dissipative. In practice, the Godunov flux based on the exact solution of the Riemann problem is rarely used, since it would require a computationally expensive solution at each quadrature node on each edge in every time-step. Thus, it is common to use approximate Riemann solvers in the context of Godunov-type methods (Toro, 2009, Chapter 9).

Convective fluxes A simple discretization of the convective term is an upwind flux formulation in the form of

$$\hat{F}\left(\psi_h^{\text{in}}, \psi_h^{\text{out}}, \mathbf{n}_\Gamma\right) = \begin{cases} \mathbf{f}(\psi_h^{\text{in}}) \cdot \mathbf{n}, & \text{if } |\mathbf{u}_\perp| > 0 \\ \mathbf{f}(\psi_h^{\text{out}}) \cdot \mathbf{n}, & \text{if } |\mathbf{u}_\perp| < 0\,, \end{cases} \tag{3.41}$$

where $|\mathbf{u}_\perp|$ denotes the flow velocity normal to an edge. In the case of linear advection, it is often chosen as $|\mathbf{u}_\perp| = \mathbf{u} \cdot \mathbf{n}_\Gamma$. The upwind flux (3.41) is very stable due to its inherently high numerical dissipation. Usually, more advanced choices for the numerical flux are employed, such as the commonly used Harten-Lax-van Leer-Compact (HLLC) flux (Toro et al., 1994) in the context of the Euler equations (2.5). An introduction to the HLLC approximate Riemann solver tailored to the *CNS* solver of the *BoSSS* framework was published by Krämer-Eis (2017). Here, only a brief summary based on Toro (2009, Chapter 10) is given.

The HLLC Riemann solver by Toro et al. (1994) is an extension of the Harten-Lax-van Leer (HLL) solver proposed by Harten et al. (1983). The basic idea of both schemes is that the required wave speeds are approximated by some independent algorithm and the integral form of the conservation laws delivers an approximated flux. The HLL scheme is based on a *two-wave* model, whereas the HLLC scheme is based on a *three-wave* model, which resolves the intermediate wave more accurately. Figure 3.3 shows the wave pattern of the HLLC scheme. The HLLC approximate Riemann solver is given by

$$\tilde{\mathbf{U}}(x_1, t) = \begin{cases} \mathbf{U}_{\text{L}}\,, & \text{if } \frac{x_1}{t} \leq \tilde{S}_{\text{L}}\,, \\ \mathbf{U}_{*\text{L}}\,, & \text{if } \tilde{S}_{\text{L}} \leq \frac{x_1}{t} \leq \tilde{S}_{*}\,, \\ \mathbf{U}_{*\text{R}}\,, & \text{if } \tilde{S}_{*} \leq \frac{x_1}{t} \leq \tilde{S}_{\text{R}}\,, \\ \mathbf{U}_{\text{R}}\,, & \text{if } \frac{x_1}{t} \geq \tilde{S}_{\text{R}}\,, \end{cases} \tag{3.42}$$

and the corresponding numerical flux is given by

$$\hat{\mathbf{F}}^{\text{HLLC}}(x_1, t) = \begin{cases} \hat{\mathbf{F}}_{\text{L}}\,, & \text{if } \frac{x_1}{t} \leq \tilde{S}_{\text{L}}\,, \\ \hat{\mathbf{F}}_{*\text{L}}\,, & \text{if } \tilde{S}_{\text{L}} \leq \frac{x_1}{t} \leq \tilde{S}_{*}\,, \\ \hat{\mathbf{F}}_{*\text{R}}\,, & \text{if } \tilde{S}_{*} \leq \frac{x_1}{t} \leq \tilde{S}_{\text{R}}\,, \\ \hat{\mathbf{F}}_{\text{R}}\,, & \text{if } \frac{x_1}{t} \geq \tilde{S}_{\text{R}}\,, \end{cases} \tag{3.43}$$

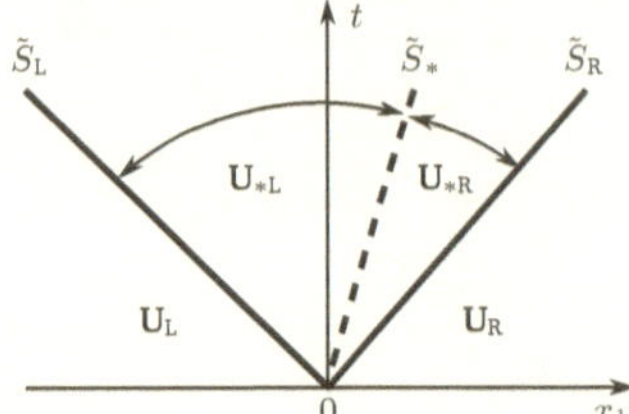

Figure 3.3: Three-wave model of the HLLC approximate Riemann solver (Toro et al., 1994). In comparison to the original HLL solver (Harten et al., 1983), the star region in the HLLC solver is additionally separated by an intermediate wave with speed $\tilde{S}_*$ into two star states $\mathbf{U}_{*L}$ and $\mathbf{U}_{*R}$ (adapted from Toro, 2009, Figure 10.4).

where the intermediate wave fluxes $\hat{\mathbf{F}}_{*L}$ and $\hat{\mathbf{F}}_{*R}$ still need to be determined. For the Euler equations, the intermediate fluxes can be calculated by using the Rankine-Hugoniot conditions, resulting in

$$\hat{\mathbf{F}}_{*L} = \hat{\mathbf{F}}_L + \tilde{S}_L \left(\mathbf{U}_{*L} - \mathbf{U}_L\right), \tag{3.44a}$$

$$\hat{\mathbf{F}}_{*R} = \hat{\mathbf{F}}_{*L} + \tilde{S}_* \left(\mathbf{U}_{*R} - \mathbf{U}_{*L}\right), \tag{3.44b}$$

$$\hat{\mathbf{F}}_{*R} = \hat{\mathbf{F}}_R + \tilde{S}_R \left(\mathbf{U}_{*R} - \mathbf{U}_R\right). \tag{3.44c}$$

So far, we have three equations (3.44) for the four vectors of unknowns $\mathbf{U}_{*L}$, $\hat{\mathbf{F}}_{*L}$, $\mathbf{U}_{*R}$, and $\hat{\mathbf{F}}_{*R}$. Note that the states $\mathbf{U}_L$ and $\mathbf{U}_R$ and the corresponding fluxes $\hat{\mathbf{F}}_L$ and $\hat{\mathbf{F}}_R$ are known. In order to solve the problem (3.44) for $\hat{\mathbf{F}}_{*L}$ and $\hat{\mathbf{F}}_{*R}$, it is sufficient to determine $\mathbf{U}_{*L}$ and $\mathbf{U}_{*R}$. Additionally, we need an expression for the still unknown intermediate wave speed $\tilde{S}_*$, which can be obtained by

$$\tilde{S}_* = \frac{p_R - p_L + \rho_L u_{1,L}(\tilde{S}_L - u_{1,L}) - \rho_R u_{1,R}(\tilde{S}_R - u_{1,R})}{\rho_L(\tilde{S}_L - u_{1,L}) - \rho_R(\tilde{S}_R - u_{1,R})}. \tag{3.45}$$

Equation (3.45) clarifies that we only need wave speed estimates for $\tilde{S}_L$ and $\tilde{S}_R$, like in the original HLL scheme. After some algebraic manipulations, we obtain the intermediate fluxes through Equations (3.44a) and (3.44c) by using the expression

$$\mathbf{U}_{*K} = \rho_K \left(\frac{\tilde{S}_K - u_{1,K}}{\tilde{S}_K - \tilde{S}_*}\right) \left[\begin{array}{c} 1 \\ \tilde{S}_* \\ \frac{\rho E_K}{\rho_K} + (\tilde{S}_* - u_{1,K})\left[\tilde{S}_* + \frac{p_K}{\rho_K(\tilde{S}_K - u_{1,K})}\right] \end{array}\right] \tag{3.46}$$

for the states in the star region, with the placeholder $K = \{L, R\}$ (Krämer-Eis, 2017). Finally, the flux value is determined by Equation (3.43). Note that the Riemann problem for higher dimensions can be broken down into the presented one-dimensional cases. Equations (3.44) to (3.46) are valid for every equation of state (EOS) as soon as estimates for the wave speeds $\tilde{S}_L$ and $\tilde{S}_R$ are given (Toro, 2009, Section 10.4). In the *BoSSS* framework, we use an estimated pressure in the star region

$$p_{pvrs} = \frac{1}{2}(p_L + p_R) - \frac{1}{2}(u_{1,R} - u_{1,L})\left(\frac{1}{2}(\rho_L + \rho_R)\frac{1}{2}(a_L + a_R)\right), \tag{3.47a}$$

$$p_* = \max(0, p_{pvrs}) \tag{3.47b}$$

in order to approximate the wave speeds

$$\tilde{S}_\mathrm{L} = u_{1,\mathrm{L}} - a_\mathrm{L}(\cdot)_\mathrm{L}\,, \qquad \tilde{S}_\mathrm{R} = u_{1,\mathrm{R}} + a_\mathrm{R}(\cdot)_\mathrm{R}\,, \tag{3.48}$$

with

$$(\cdot)_\mathrm{K} = \begin{cases} 1\,, & \text{if } p_* \le p_\mathrm{K}\,, \\ \left[1 + \frac{\gamma+1}{2\gamma}\left(\frac{p_*}{p_\mathrm{K}} - 1\right)\right]^{1/2}\,, & \text{if } p_* > p_\mathrm{K}\,, \end{cases} \tag{3.49}$$

for $\mathrm{K} = \{\mathrm{L},\mathrm{R}\}$ (Toro, 2009, Section 10.6). Then, the intermediate wave speed can be calculated by inserting Equations (3.47) to (3.49) into Equation (3.45).

Diffusive fluxes We apply the symmetric interior penalty (SIP) method (Arnold, 1982) for the discretization of second-order diffusive terms. In Section 5.3, we will extend the Euler equations (2.5) by an artificial viscosity term in the context of a two-step shock-capturing strategy to

$$\underbrace{\frac{\partial \mathbf{U}}{\partial t} + \frac{\partial \mathbf{F}_1(\mathbf{U})}{\partial x_1} + \frac{\partial \mathbf{F}_2(\mathbf{U})}{\partial x_2}}_{\text{Euler equations (2.5)}} = \underbrace{\varepsilon\left(\frac{\partial^2 \mathbf{U}}{\partial x_1^2} + \frac{\partial^2 \mathbf{U}}{\partial x_2^2}\right)}_{\substack{\text{artificial viscosity} \\ \text{term}}}. \tag{3.50}$$

The SIP method can be written in the bilinear form

$$\hat{a}_\mathrm{SIP}(\mathbf{u},\mathbf{v}) = \int_\Omega \underbrace{\mu(\nabla_h\mathbf{u} \cdot \nabla_h\mathbf{v})}_{\text{volume term}} \mathrm{d}V - \oint_{\Gamma\backslash\Gamma_\mathrm{Neu}} \underbrace{\{\!\{\mu\nabla_h\mathbf{u}\}\!\} \cdot \mathbf{n}_\Gamma \, [\![\mathbf{v}]\!]}_{\text{consistency term}} + \underbrace{\{\!\{\mu\nabla_h\mathbf{v}\}\!\} \cdot \mathbf{n}_\Gamma \, [\![\mathbf{u}]\!]}_{\text{symmetry term}} \mathrm{d}S$$

$$+ \oint_{\Gamma\backslash\Gamma_\mathrm{Neu}} \underbrace{\eta\, [\![\mathbf{u}]\!]\, [\![\mathbf{v}]\!]}_{\text{penalty term}} \mathrm{d}S\,, \tag{3.51}$$

with a viscosity parameter μ, being a stable and consistent discretization of a Poisson-like model problem with homogeneous Dirichlet boundary conditions

$$-\Delta(\mu\mathbf{u}) = \mathbf{g}_\Omega\,, \qquad \text{in } \Omega\,, \tag{3.52a}$$
$$\mathbf{u} = \mathbf{g}_D = 0\,, \qquad \text{on } \partial\Omega = \Gamma_\mathrm{Dir} \tag{3.52b}$$

in the Sobolev space

$$\hat{H}_0^1 := \left\{\mathbf{u} \in L^2(\Omega)\,; \|\mathbf{u}\|_2 + \|\nabla\mathbf{u}\|_2 < \infty\,; \mathbf{u}|_{\partial\Omega} = 0\right\}\,, \tag{3.53}$$

following Kummer and Müller (2020). Now, we look for a weak solution $\mathbf{u} \in \hat{H}_0^1(\Omega)$ such that

$$\hat{a}(\mathbf{u},\mathbf{v}) = \hat{a}_\mathrm{SIP}(\mathbf{u},\mathbf{v}) = \int_\Omega \mathbf{g}_\Omega\mathbf{v}\,\mathrm{d}V\,, \qquad \mathbf{v} \in \hat{H}_0^1(\Omega)\,. \tag{3.54}$$

The penalty parameter η in Equation (3.51) has to be chosen large enough to ensure coercivity of the bilinear form (3.51). However, it should not be chosen arbitrarily large in order to avoid over-penalization. This would increase the condition number of the operator matrix and, thus, the approximation error. Additionally, it decreases the maximum stable time-step size in the context of explicit time-integration schemes. We choose the cell-wise constant penalty parameter, omitting the cell index j, as

$$\eta := \max\left(\mu^{\text{in}}, \mu^{\text{out}}\right) \cdot \max\left(\tilde{\eta}^{\text{in}}, \tilde{\eta}^{\text{out}}\right), \qquad \text{on } \Gamma_{\text{int}}, \tag{3.55a}$$

$$\eta := \mu^{\text{in}} \cdot \tilde{\eta}^{\text{in}}, \qquad \text{on } \partial\Omega, \tag{3.55b}$$

where we introduce a cell-local penalty factor

$$\tilde{\eta} = \frac{\eta_0 P^2}{h}. \tag{3.56}$$

The cell-local penalty factor (3.56) depends on a safety factor η_0, which is set to $\eta_0 = 4.0$, on the polynomial degree P and on a characteristic length scale h. Further details about the choice of the penalty parameter are given by Kummer (2016) and Müller et al. (2017).

Cell-local characteristic length scale A cell-local characteristic length scale h_j is required to determine the penalty parameters (3.55) and (3.56). Thus, we introduce a definition of h_j in the boundary fitted context and also in the XDG context, see also Sections 3.3 and 4.1.3, as

$$h_j := \begin{cases} \min\left(|\mathcal{E}_{j,e}|\right), & \text{if } K_j \text{ is a standard or agglomeration target cell}, \\ |K_{j,\mathfrak{s}}| \,/\, |\partial K_j|, & \text{if } K_j \text{ is a non-agglomerated cut-cell}, \end{cases} \tag{3.57}$$

where $\mathcal{E}_{j,e}$ with $e = 1, \ldots, E$ denotes an edge of a cell K_j and $\mathfrak{s}$ denotes the active species in the context of the XDG method. Note that in the applied XDG method h_j is already known from the creation of the quadrature rules and does not have to be additionally calculated. For square cells, both terms in Equation (3.57) lead to identical results.

3.2.4 Boundary Conditions

We apply the boundary conditions in a weak sense by introducing *ghost states* $\mathbf{U}^{bc}$ for the values outside of the computational domain for the evaluation of the numerical fluxes (Krämer-Eis, 2017). For details about the treatment of boundary conditions for compressible flow in the context of the DG method, the reader is referred to the works by Bassi and Rebay (1997), Hartmann and Houston (2008), and Mengaldo et al. (2014). The characteristics and their speeds, being $u_1 - a$, u_1, and $u_1 + a$ in the quasi-one-dimensional case, strongly influence the type of the boundary condition, see Equations (3.31), (3.33) and (3.34).

We define the ghost states $\mathbf{U}^{bc}$ in two dimensions by using the free-stream conditions $(\cdot)_\infty$ or the inner values of the boundary cells $(\cdot)^{\text{in}}$:

- *Supersonic inlet:* All three characteristics ($u_1 - a > 0, u_1 > 0, u_1 + a > 0$) enter the domain. Thus, we apply the free-stream conditions

$$\mathbf{U}^{bc}_{\text{in,super}} := \begin{pmatrix} \rho_\infty \\ \rho_\infty u_{\infty,1} \\ \rho_\infty u_{\infty,2} \\ \rho E_\infty \end{pmatrix}. \tag{3.58}$$

- *Subsonic inlet:* Two characteristics ($u_1 > 0, u_1 + a > 0$) enter the domain, one characteristic ($u_1 - a < 0$) leaves the domain. We use the free-stream conditions except for the pressure where we use the inner value

$$\mathbf{U}^{bc}_{in,sub} := \begin{pmatrix} \rho_\infty \\ \rho_\infty u_{\infty,1} \\ \rho_\infty u_{\infty,2} \\ \frac{p^{in}}{\gamma-1} + \rho_\infty \mathbf{u}_\infty \cdot \mathbf{u}_\infty \end{pmatrix}. \tag{3.59}$$

- *Supersonic outlet:* All characteristics ($u_1 - a > 0, u_1 > 0, u_1 + a > 0$) leave the domain. Thus, we use the inner values

$$\mathbf{U}^{bc}_{out,super} := \begin{pmatrix} \rho^{in} \\ (\rho u_1)^{in} \\ (\rho u_2)^{in} \\ (\rho E)^{in} \end{pmatrix}. \tag{3.60}$$

- *Subsonic outlet:* Two characteristics ($u_1 > 0, u_1 + a > 0$) leave the domain and one characteristic ($u_1 - a < 0$) enters the domain. We use the inner values except for the pressure

$$\mathbf{U}^{bc}_{out,sub} := \begin{pmatrix} \rho^{in} \\ (\rho u_1)^{in} \\ (\rho u_2)^{in} \\ \frac{p^{out}}{\gamma-1} + \rho^{in} \mathbf{u}^{in}_\infty \cdot \mathbf{u}^{in}_\infty \end{pmatrix}. \tag{3.61}$$

- *Adiabatic slip-wall:* In the context of the Euler equations (2.5) it is only possible to define a wall boundary condition where the normal velocity vanishes ($\mathbf{u} \cdot \mathbf{n}_\Gamma = 0$) and a tangential velocity is allowed. We achieve this by introducing a *mirrored state* of the momentum vector in normal direction $\mathbf{m}^{bc} = \mathbf{m}^{in} - 2\rho^{in}(\mathbf{u}^{in} \cdot \mathbf{n}_\Gamma)\mathbf{n}_\Gamma$, while still preserving the properties of the characteristics (Van der Vegt and Van der Ven, 2002). This is beneficial for the implementation of the HLLC fluxes, since the same function can be evaluated in the bulk and at the wall. The adiabatic slip-wall boundary condition is given by

$$\mathbf{U}^{bc}_{wall} := \begin{pmatrix} \rho^{in} \\ (\rho u_1)^{bc} \\ (\rho u_2)^{bc} \\ \frac{p^{in}}{\gamma-1} + \rho^{in} \mathbf{u}^{bc}_\infty \cdot \mathbf{u}^{bc}_\infty \end{pmatrix}. \tag{3.62}$$

- *Artificial viscosity:* We discretize the artificial viscosity term, appearing in the concept of shock-capturing in Section 5.3, by an SIP flux formulation (3.51). We use zero-Neumann boundary conditions for the artificial viscosity term at the domain boundaries

$$\mathbf{U}^{in} \cdot \mathbf{n}_\Gamma = 0. \tag{3.63}$$

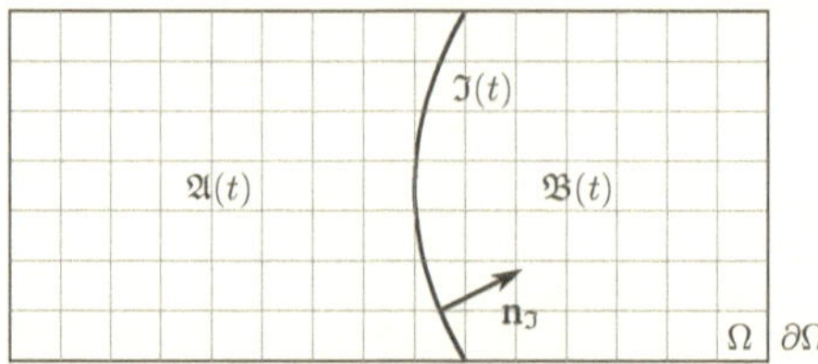

Figure 3.4: Sharp interface description in the context of an XDG method. The computational domain Ω is partitioned into two disjoint regions, denoted by the species $\mathfrak{A}(t)$ and $\mathfrak{B}(t)$, which are separated by the interface $\mathfrak{I}(t) := \overline{\mathfrak{A}}(t) \cap \overline{\mathfrak{B}}(t)$ with its normal vector $\mathbf{n}_\mathfrak{I} = \nabla\varphi/|\nabla\varphi|$. The interface is implicitly defined by the zero iso-contour of a sufficiently smooth level-set function $\varphi(\mathbf{x}, t)$.

3.3 An Extended Discontinuous Galerkin Method

In this section, we extend the generic DG method introduced in Section 3.2 to an XDG method. We introduce a sharp interface description in order to account for computational domains of arbitrary shape. Examples for sharp interfaces are the boundary of blunt bodies, shock waves or a droplet in the context of multi-phase flows. The following derivation is based on the works by Kummer (2016), Geisenhofer et al. (2019), Kummer et al. (2020), and Smuda (2020).

Sharp interface description   In the *BoSSS* framework, a sharp interface is implicitly described by means of the zero iso-contour of a sufficiently smooth level-set function $\varphi(\mathbf{x}, t)$. In Definition 3.1, we have introduced the computational domain $\Omega \subset \mathbb{R}^2$, being of polygonal shape and consisting of a set of non-overlapping cells $\mathcal{K}_h$. As illustrated in Figure 3.4, we now partition the computational domain Ω into two disjoint regions, also called *species*, $\mathfrak{A}(t)$ and $\mathfrak{B}(t)$, separated by the interface $\mathfrak{I}(t) := \overline{\mathfrak{A}}(t) \cap \overline{\mathfrak{B}}(t)$, which is assumed to be a $(D-1)$-dimensional manifold so that

$$\Omega = \mathfrak{A}(t) \;\dot{\cup}\; \mathfrak{I}(t) \;\dot{\cup}\; \mathfrak{B}(t) \,. \tag{3.64}$$

The partitioning (3.64) is implicitly given by the level-set function $\varphi(\mathbf{x}, t)$ as

$$\mathfrak{A}(t) = \{\mathbf{x} \in \Omega : \varphi(\mathbf{x}, t) < 0\} \,, \tag{3.65a}$$
$$\mathfrak{I}(t) = \{\mathbf{x} \in \Omega : \varphi(\mathbf{x}, t) = 0\} \,, \tag{3.65b}$$
$$\mathfrak{B}(t) = \{\mathbf{x} \in \Omega : \varphi(\mathbf{x}, t) > 0\} \,. \tag{3.65c}$$

We define the normal vector $\mathbf{n}_\mathfrak{I}$ at the interface, pointing from $\mathfrak{A}$ to $\mathfrak{B}$, by means of the normalized gradient of the level-set function and the level-set curvature $\tilde{\kappa}$ using Bonnet's formula as

$$\mathbf{n}_\mathfrak{I} = \frac{\nabla\varphi}{|\nabla\varphi|} \,, \qquad \tilde{\kappa} = \nabla \cdot \left(\frac{\nabla\varphi}{|\nabla\varphi|} \right) \,. \tag{3.66}$$

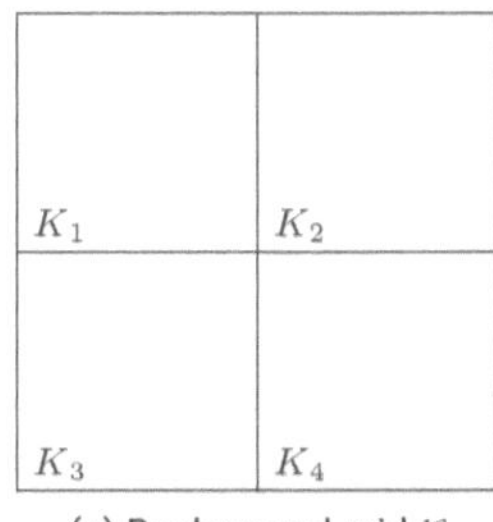

(a) Background grid $\mathcal{K}_h$.

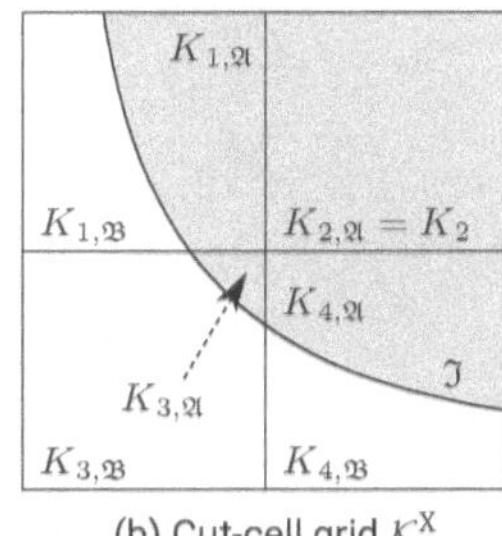

(b) Cut-cell grid $\mathcal{K}_h^{\mathsf{X}}$.

Figure 3.5: Cut-cell grid with its corresponding background grid. The interface $\mathfrak{J}$ in combination with the species $\mathfrak{A}$ (gray) and the species $\mathfrak{B}$ (white) build a cut-cell grid $\mathcal{K}_h^{\mathsf{X}} = \{K_{1,\mathfrak{A}}, K_{1,\mathfrak{B}}, K_{2,\mathfrak{A}} = K_2, K_{2,\mathfrak{B}} = \emptyset, K_{3,\mathfrak{A}}, K_{3,\mathfrak{B}}, K_{4,\mathfrak{A}}, K_{4,\mathfrak{B}}\}$ based on the background grid $\mathcal{K}_h = \{K_1, K_2, K_3, K_4\}$ (adapted from Kummer et al., 2020, Figure 1).

3.3.1 Extended Definitions

We extend the definitions of a generic DG method, see Section 3.2.1, in order to incorporate the sharp interface description in the context of the XDG method. By introducing the interface $\mathfrak{J}$, we obtain a discrete set of *cut-cells* with a species $\mathfrak{s}(t) \in \{\mathfrak{A}(t), \mathfrak{B}(t)\}$ being 'active' on one side of the interface or on the other, respectively. Figure 3.5 shows an exemplary cut-cell grid.

Definition 3.4 (Cut-cells and cut-cell grid). Time-dependent cut-cells are defined as

$$K_{j,\mathfrak{s}}(t) := K_j \cap \mathfrak{s}(t) \tag{3.67}$$

for a species $\mathfrak{s} \in \{\mathfrak{A}(t), \mathfrak{B}(t)\}$. The set of all cut-cells $K_{j,\mathfrak{s}}(t)$ forms the cut-cell grid

$$\mathcal{K}_h^{\mathsf{X}}(t) = \{K_{1,\mathfrak{A}}, K_{1,\mathfrak{B}}, \ldots, K_{J,\mathfrak{A}}, K_{J,\mathfrak{B}}\}. \tag{3.68}$$

From now on we will refer to the cells K_j of the generic DG method, see Definition 3.1, as *background cells* in contrast to the newly introduced cut-cells $K_{j,\mathfrak{s}}$.

Definition 3.5 (Extended notations). The two-dimensional polygonal and simply connected domain now reads $\Omega = \mathfrak{A}(t) \cup \mathfrak{J}(t) \cup \mathfrak{B}(t) \subset \mathbb{R}^2$. On this domain, we define the following quantities:

- The computational cut-cell grid $\mathcal{K}_h^{\mathsf{X}}(t)$, as given by Equation (3.68).

- The set of all edges Γ is extended to $\Gamma := \cup_j \partial K_j \cup \mathfrak{J}$.

- The normal field $\mathbf{n}_\Gamma$ on Γ also incorporates the interface normal $\mathbf{n}_\Gamma = \mathbf{n}_\mathfrak{J}$.

The XDG method can be seen as a DG method on a cut-cell grid, which motivates the definition of the XDG space in Definition 3.6 (Kummer et al., 2020).

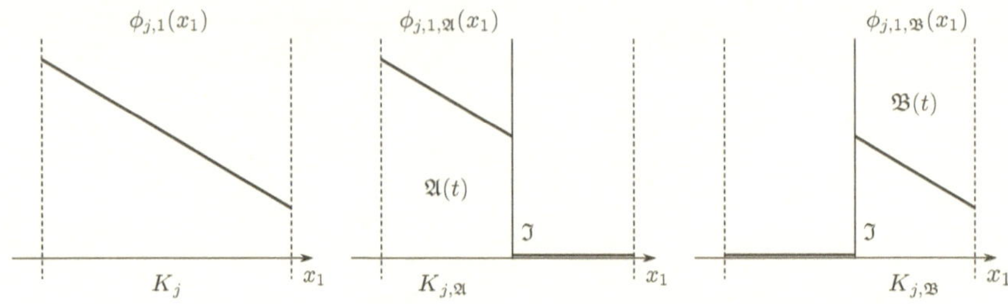

(a) Background cell $K_j \in \mathcal{K}_h$.   (b) Cut-cell for species $\mathfrak{A}(t)$ $(K_{j,\mathfrak{A}} = K_j \cap \mathfrak{A}(t))$.   (c) Cut-cell for species $\mathfrak{B}(t)$ $(K_{j,\mathfrak{B}} = K_j \cap \mathfrak{B}(t))$.

Figure 3.6: Linear polynomial basis functions in a one-dimensional cut-cell. (a) The basis function $\phi_{j,1}(x_1)$ is defined in the entire background cell. (b) and (c) The basis function $\phi_{j,1}(x_1)$ is cut at the interface $\mathfrak{J}$, resulting in $\phi_{j,1,\mathfrak{A}}(x_1) = \phi_{j,1}(x_1)\mathbb{1}_{\mathfrak{A}}(t,x_1)$ and $\phi_{j,1,\mathfrak{B}}(x_1) = \phi_{j,1}(x_1)\mathbb{1}_{\mathfrak{B}}(t,x_1)$ for species $\mathfrak{A}(t)$ and $\mathfrak{B}(t)$, respectively, as given in Equation (3.70). In $K_{j,\mathfrak{A}}$ and in $K_{j,\mathfrak{B}}$, there is a separated set of DOF.

Definition 3.6 (Extended discontinuous Galerkin space). We define the broken polynomial cut-cell XDG space

$$\mathbb{P}_P^X(\mathcal{K}_h, t) := \left\{ f \in L^2(\Omega); \forall K \in \mathcal{K}_h : f|_{K \cap \mathfrak{s}(t)} \text{ is polynomial and } \deg(f|_{K \cap \mathfrak{s}(t)}) \leq P \right\} \tag{3.69}$$
$$= \mathbb{P}_P(\mathcal{K}_h^X(t))$$

with a total polynomial degree of P. If there is only one species $\mathfrak{s}(t)$ present in a specific background cell K_j such that $K_{j,\mathfrak{s}} = K_j$, the standard DG space (3.1) in Definition 3.2 is recovered. Furthermore, the cell-local approximation (3.7) is valid. It has to be adapted for cases when two species $\mathfrak{s}(t) \in \{\mathfrak{A}(t), \mathfrak{B}(t)\}$ are present. In this case, the scalar property $\psi_j(\mathbf{x}, t)$ in a cell $K_j \in \mathcal{K}_h$ is approximated by

$$\psi_j(\mathbf{x}, t) = \sum_{n=1}^{N} \tilde{\psi}_{j,n,\mathfrak{A}}(t) \underbrace{\phi_{j,n}(\mathbf{x})\mathbb{1}_{\mathfrak{A}}(\mathbf{x}, t)}_{\phi_{j,n,\mathfrak{A}}(\mathbf{x},t)} + \sum_{n=1}^{N} \tilde{\psi}_{j,n,\mathfrak{B}}(t) \underbrace{\phi_{j,n}(\mathbf{x})\mathbb{1}_{\mathfrak{B}}(\mathbf{x}, t)}_{\phi_{j,n,\mathfrak{B}}(\mathbf{x},t)}, \qquad \mathbf{x} \in K_j, \tag{3.70}$$

where $\tilde{\psi}_{j,n,\mathfrak{s}}(t)$ denote the unknown coefficients in species $\mathfrak{s}(t)$, $\mathbb{1}_{\mathfrak{s}}(\mathbf{x}, t)$ denotes the characteristic function for species $\mathfrak{s}(t)$ ($\mathbb{1}_{\mathfrak{s}}(\mathbf{x}, t) = 1$ for $\mathbf{x} \in \mathfrak{s}(t)$, and $\mathbb{1}_{\mathfrak{s}}(\mathbf{x}, t) = 0$ for $\mathbf{x} \notin \mathfrak{s}(t)$) and $\phi_{j,n,\mathfrak{s}}(\mathbf{x}, t)$ denote the polynomial basis functions for species $\mathfrak{s}(t)$, which are cut at the interface $\mathfrak{J}$. The polynomial basis functions for both the DG method and the XDG method are identical. For illustration purposes, Figure 3.6 shows the cut polynomial basis functions for a linear basis in a one-dimensional cut background cell K_j consisting of $K_{j,\mathfrak{A}}$ and $K_{j,\mathfrak{B}}$. Note that there is a separated set of DOF in each species $\mathfrak{s}(t)$.

3.3.2 Cell-Agglomeration

Arbitrarily shaped cut scenarios can arise depending on the position and on the shape of the interface $\mathfrak{J}$ in Figure 3.4. Small and ill-shaped cut-cells are challenging issues for numerical

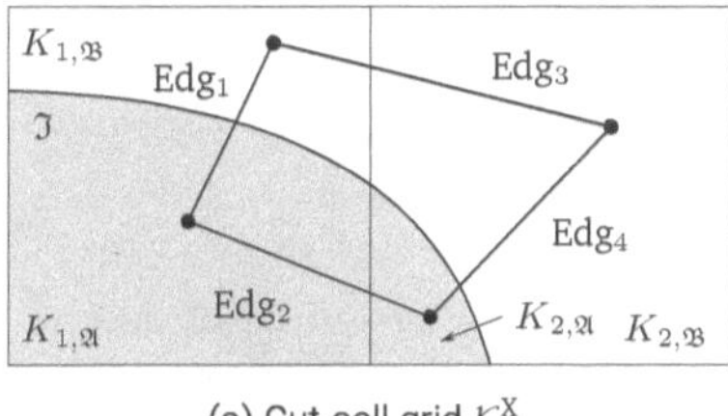

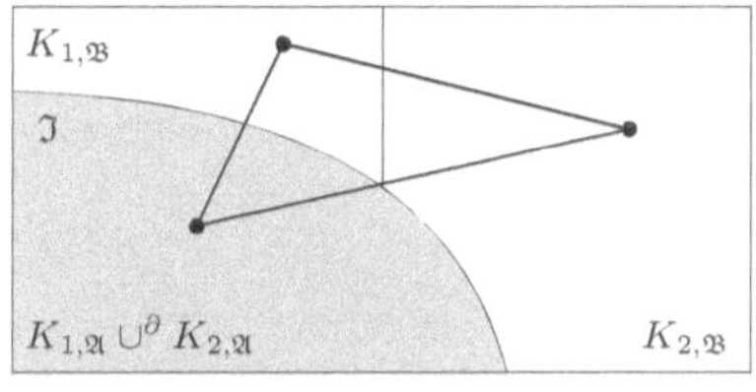

(a) Cut-cell grid $\mathcal{K}_h^X$. (b) Agglomeration grid $\mathrm{Agg}(\mathcal{K}_h^X, A_{\tilde{a}})$.

Figure 3.7: Cell-agglomeration by means of an undirected graph $\mathrm{Gr}(\mathcal{K}_h^X)$. The undirected graph $\mathrm{Gr}(\mathcal{K}_h^X) = (\mathcal{K}_h^X, \mathrm{Edg}(\mathcal{K}_h^X))$ consists of the set of edges $\{\mathrm{Edg}_1(\{K_{1,\mathfrak{A}}, K_{1,\mathfrak{B}}\}),$ $\mathrm{Edg}_2(\{K_{1,\mathfrak{A}}, K_{2,\mathfrak{A}}\}), \mathrm{Edg}_3(\{K_{1,\mathfrak{B}}, K_{2,\mathfrak{B}}\}), \mathrm{Edg}_4(\{K_{2,\mathfrak{A}}, K_{2,\mathfrak{B}}\})\}$ on the cut-cell grid $\mathcal{K}_h^X$. The corresponding agglomeration grid $\mathrm{Agg}(\mathcal{K}_h^X, A_{\tilde{a}})$ contains the agglomeration map $A_{\tilde{a}} = \{\mathrm{Edg}_2(\{K_{1,\mathfrak{A}}, K_{2,\mathfrak{A}}\})\}$. Note that only agglomeration across the same species is allowed (adapted from Smuda, 2020, Figure 3.4).

integration schemes and in particular for explicit time-integration schemes, due to the increasing condition numbers of the cell-local mass matrices, or due to the prohibitive CFL-condition, see Section 4.1.3. A remedy for these issues are cell-agglomeration techniques, where undesired cut-cells $K_{j,\mathfrak{s}}$ are removed from the cut-cell grid $\mathcal{K}_h^X$. Then, all operations are carried out on an agglomerated grid $\mathcal{K}_h^{X,\mathrm{agg}}$, see Definition 3.8, before the solution is projected back onto the cut-cell grid $\mathcal{K}_h^X$. We follow the work by Kummer et al. (2020) and introduce the cell-agglomeration technique by means of graph theory.

Definition 3.7 (Graph of a numerical grid). We consider a numerical grid $\mathcal{K}$, which can be the background grid $\mathcal{K}_h$ (Definition 3.1) or the cut-cell grid $\mathcal{K}_h^X$ (Definition 3.4), for example.

When considering two cells $K_j, K_l \in \mathcal{K}$, we define a *logical edge* $\mathrm{Edg}(\{K_j, K_l\})$ by using the set $\{K_j, K_l\}$, if and only if, both cells share a common edge $(\oint_{\overline{K_j} \cap \overline{K_l}} 1 \, dS > 0)$. We denote the set of all logical edges with $\mathrm{Edg}(\mathcal{K})$. In terms of graph theory, we define an *undirected graph* as $\mathrm{Gr}(\mathcal{K}_h) := (\mathcal{K}_h, \mathrm{Edg}(\mathcal{K}_h))$.

Definition 3.8 (Agglomeration maps and grids). We consider a subset of edges $\tilde{A} \subset \mathrm{Edg}(\mathcal{K}_h)$ of the computational grid $\mathcal{K}_h$. Next, we define a cluster of cells $\tilde{a} := \{K_1, \ldots, K_L\} \in \mathcal{K}_h$. The cluster of cells $\tilde{a}$ represents the maximum number of cells, which are connected by the subset of edges $\tilde{A} \subset \mathrm{Edg}(\mathcal{K}_h)$. In particular, we introduce a so-called *agglomeration map* linked with the cluster $\tilde{a}$ by the notation $A_{\tilde{a}} \subset \mathrm{Edg}(\mathcal{K}_h)$ with the subscript $\tilde{a}$. We define the agglomeration of all cells $K_l \in \tilde{a}$ as $\mathcal{K}_{\tilde{a}} := \cup^\partial_{K_l \in \tilde{a}} K_l$. Here, we use a modified union $X \cup^\partial Y := (\overline{X} \cup \overline{Y}) \setminus \partial(\overline{X} \cup \overline{Y})$ in order to guarantee that the agglomeration of the cells $\mathcal{K}_{\tilde{a}}$ is still a simply connected, open set by taking the closure of each cell and then by subtracting its boundary. The cluster $\tilde{a}$ can consist of a single cell which is then called a non-agglomerated cell with respect to $\tilde{a}$.

All agglomerated and non-agglomerated cells with their corresponding agglomeration map $A_{\tilde{a}} \subset \mathrm{Edg}(\mathcal{K}_h)$ define the *agglomeration grid* $\mathrm{Agg}(\mathcal{K}_h, A_{\tilde{a}}) := \mathcal{K}_h^{X,\mathrm{agg}}$. Figure 3.7 shows the agglomeration procedure on an exemplary cut-cell grid. This motivates the definition of the agglomerated cut-cell XDG space in Definition 3.9.

Definition 3.9 (Agglomerated extended discontinuous Galerkin (XDG) space). On a given agglomeration grid $\mathrm{Agg}(\mathcal{K}_h^X, A_{\tilde{a}})$ of the cut-cell grid $\mathcal{K}_h^X$ with the corresponding agglomeration map $A_{\tilde{a}} \subset \mathrm{Edg}(\mathcal{K}_h^X)$, we define the broken polynomial agglomerated cut-cell XDG space as

$$\mathbb{P}_P^{X,A_{\tilde{a}}} := \mathbb{P}_P(\mathrm{Agg}(\mathcal{K}_h^X, A_{\tilde{a}})) , \tag{3.71}$$

which is a sub-space of the original XDG space $\mathbb{P}_P(\mathcal{K}_h^X)$. The reader is referred to Equations (3.1) and (3.69) for the definition of the standard DG space and the cut-cell XDG space, respectively.

Subsequently, we introduce the volume fraction $|K_{j,\mathfrak{s}}| / |K_j|$ of a cut-cell $K_{j,\mathfrak{s}}$ with respect to the corresponding background cell K_j as a measure for the size of cut-cells. As previously mentioned, the cell-agglomeration technique removes undesired cut-cells from $\mathcal{K}_h^X$. Consequently, a valid agglomeration map $A_{\tilde{a}} \subset \mathrm{Edg}(\mathcal{K}_h^X)$ must fulfill the following two properties:

(1) All cut-cells with a volume fraction smaller than an agglomeration threshold δ_{agg},

$$0 \le \frac{|K_{j,\mathfrak{s}}|}{|K_j|} \le \delta_{\mathrm{agg}} , \tag{3.72}$$

are agglomerated to a neighboring target cell (see (2)). Typically, the agglomeration threshold is chosen in the range of $0.1 \le \delta_{\mathrm{agg}} \le 0.3$. On the one hand, this leads to reasonable condition numbers of the mass matrices and additionally allows for larger time-step sizes when using explicit time-integration schemes. On the other hand, the discretization errors near the interface are still sufficiently small (Kummer, 2016; Müller et al., 2017).

(2) The agglomeration target cell is the neighboring cell with the largest volume fraction of the same species $\mathfrak{s}$. Consequently, no agglomeration across different species is allowed so that in the case of two species $\mathfrak{A}$ and $\mathfrak{B}$ there is no edge $\{K_{j,\mathfrak{A}}, K_{j,\mathfrak{B}}\}$ in the agglomeration map $A_{\tilde{a}}$.

Algorithm 3.1 describes the procedure of how to find the cut-cells which will be agglomerated (*source cells*) and how to find the corresponding neighboring cells (*target cells*) which will 'receive' the source cells. In particular, Algorithm 3.1 presents the construction of an agglomeration map $A_{\tilde{a}} = FindAgg(\mathcal{K}_h^X, \delta_{\mathrm{agg}})$ depending on an agglomeration threshold δ_{agg}. Target cells may become source cells themselves depending on the form of the interface leading to so-called agglomeration chains. In general, the presented Algorithm 3.1 can handle such scenarios, but we did not encounter any relevant configuration in the test cases of this work.

Implementation   Notes about the implementation and the computational effort of the agglomeration procedures in the context of compressible flows are given in the work by Müller et al. (2017). Here, we want to introduce the concept briefly.

In the following, we denote a cut-cell $K_{j,\mathfrak{s}}$ which will be agglomerated to a cell of the same species as the *source cell* K^{src}, and the corresponding neighboring cell $K_{k,\mathfrak{s}}$ with the largest volume fraction of the same species as the *target cell* K^{tar}. Figure 3.8 shows an exemplary cut configuration on the background grid $\mathcal{K}_h$ and on the agglomerated grid $\mathcal{K}_h^{X,\mathrm{agg}}$, respectively. We denote the row vectors of basis functions, which are only non-zero in the corresponding cell, as $\phi^{\mathrm{src}} = (\phi_1^{\mathrm{src}}, \dots, \phi_N^{\mathrm{src}})$ and $\phi^{\mathrm{tar}} = (\phi_1^{\mathrm{tar}}, \dots, \phi_N^{\mathrm{tar}})$. We skip the cell index j in this paragraph

Algorithm 3.1: Pseudocode of the presented cell-agglomeration technique. We construct an agglomeration map $A_{\tilde{a}} = \text{FindAgg}(\mathcal{K}_h^{\mathsf{X}}, \delta_{\text{agg}})$ on a given cut-cell grid $\mathcal{K}_h^{\mathsf{X}}$ for an agglomeration threshold $\delta_{\text{agg}} \in [0, 1)$ (adapted from Kummer et al., 2020, Algorithm 7).

1: **function** FINDAGG($\mathcal{K}_h^{\mathsf{X}}, \delta_{\text{agg}}$)
2:      Initiate $A_{\tilde{a}} \leftarrow \{\}$
3:      **for all** $K_j \in \mathcal{K}_h^{\mathsf{X}}$ **do**   ▷ Loop over all cells
4:          **for all** $\mathfrak{s} \in \{\mathfrak{A}, \mathfrak{B}\}$ **do**   ▷ Loop over all species
5:              **if** $|K_{j,\mathfrak{s}}| \,/\, |K_j| \leq \delta_{\text{agg}}$ **then**   ▷ Find small cut-cells
6:                  $K_{k,\mathfrak{s}} \leftarrow \underset{K_{l,\mathfrak{s}} \text{ neighb. of } K_{j,\mathfrak{s}}}{\arg\max} (|K_{l,\mathfrak{s}}| \,/\, |K_l|)$ ▷ Find largest neighbor
7:                  **if** $K_{k,\mathfrak{s}}$ exists **then**
8:                      $A_{\tilde{a}} \leftarrow A_{\tilde{a}} \cup \{\{K_{j,\mathfrak{s}}, K_{k,\mathfrak{s}}\}\}$ ▷ Add edge to agglomeration map
9:                  **end if**
10:              **end if**
11:          **end for**
12:      **end for**
13:      **return** $A_{\tilde{a}}$
14: **end function**

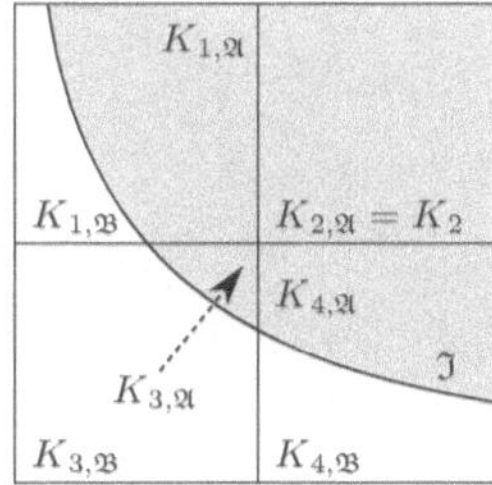

(a) Initial configuration on the non-agglomerated grid $\mathcal{K}_h^{\mathsf{X}}$.

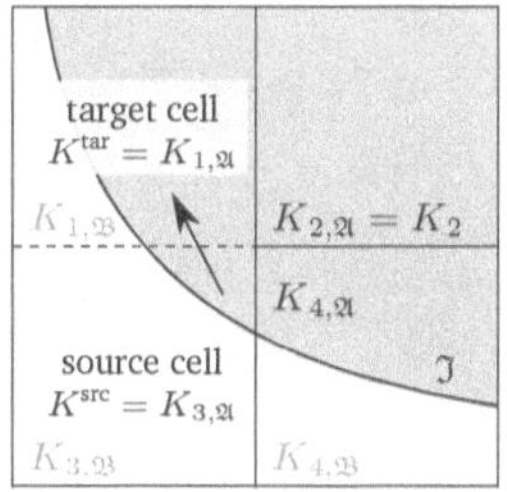

(b) Agglomerated grid $\mathcal{K}_h^{\mathsf{X,agg}}$.

Figure 3.8: Illustration of the cell-agglomeration technique. (a) Small cut-cells will be agglomerated to their direct neighbor with the largest fluid volume fraction. (b) For small values of the agglomeration threshold δ_{agg}, the source cell $K^{\text{src}} = K_{3,\mathfrak{A}}$ is agglomerated to the target cell $K^{\text{tar}} = K_{1,\mathfrak{A}}$ (adapted from Geisenhofer et al., 2019, Figure 2).

for better readability. Next, we introduce a coupling matrix $\mathbf{Q} \subset \mathbb{R}^{N,N}$, with $\phi^{\mathrm{src}}\mathbf{Q}$ being the smooth extension of ϕ^{tar} into the source cell K^{src}. Consequently, we can write the row vector of basis functions in an agglomerated cell K^{agg} as

$$\phi^{\mathrm{agg}} = \phi^{\mathrm{tar}} + \phi^{\mathrm{src}}\mathbf{Q}. \tag{3.73}$$

For fixed level-set positions, the coupling matrix $\mathbf{Q}$ has to be calculated only once in the beginning of every simulation. This is done by projecting the smooth extension of the basis of the target cell denoted by $\hat{\phi}^{\mathrm{tar}} = (\hat{\phi}_1^{\mathrm{tar}}, \ldots, \hat{\phi}_N^{\mathrm{tar}})$ onto the source cell K^{src} by calculating

$$\mathbf{Q} = \begin{pmatrix} \int_{K^{\mathrm{src}}} \hat{\phi}_1^{\mathrm{tar}} \phi_1^{\mathrm{src}} \, \mathrm{d}V & \cdots & \int_{K^{\mathrm{src}}} \hat{\phi}_N^{\mathrm{tar}} \phi_1^{\mathrm{src}} \, \mathrm{d}V \\ \vdots & \ddots & \vdots \\ \int_{K^{\mathrm{src}}} \hat{\phi}_1^{\mathrm{tar}} \phi_N^{\mathrm{src}} \, \mathrm{d}V & \cdots & \int_{K^{\mathrm{src}}} \hat{\phi}_N^{\mathrm{tar}} \phi_N^{\mathrm{src}} \, \mathrm{d}V \end{pmatrix}, \tag{3.74}$$

which directly yields the expression

$$\hat{\phi}^{\mathrm{tar}} = \phi^{\mathrm{src}}\mathbf{Q}. \tag{3.75}$$

There are two main advantages of the presented formulation of the agglomerated basis functions (3.73): First, the existing implementation, including all operators and data structures, can be reused; second, the coupling matrix $\mathbf{Q}$, defined in Equation (3.74), simplifies the (de-)agglomeration in unsteady calculations with moving interfaces. In the case of a fixed interface position, the following quantities have to be calculated before and during a steady simulation run with a fixed interface position:

- *Before the simulation run (set-up phase):* We calculate the constant agglomerated cell-local mass matrix via

$$\mathbf{M}^{\mathrm{agg}} = \mathbf{M}^{\mathrm{tar}} + \mathbf{Q}^{\mathsf{T}}\mathbf{M}^{\mathrm{src}}\mathbf{Q} \tag{3.76}$$

in order to project the initial condition onto an agglomerated cell K^{agg} by calculating the agglomerated coefficients $\tilde{\psi}^{\mathrm{agg}}$ via

$$\tilde{\psi}^{\mathrm{agg}} = (\mathbf{M}^{\mathrm{agg}})^{-1} \left(\mathbf{M}^{\mathrm{tar}}\tilde{\psi}^{\mathrm{tar}} + \mathbf{Q}^{\mathsf{T}}\mathbf{M}^{\mathrm{src}}\tilde{\psi}^{\mathrm{src}} \right). \tag{3.77}$$

After the initial projection, the solution in an agglomerated cell K^{agg} stays continuous for the entire simulation run. Thus, the problem formulation is not altered by injecting the solution $\tilde{\psi}^{\mathrm{agg}}$ from the agglomerated grid $\mathcal{K}_h^{\mathrm{X,agg}}$ onto the original non-agglomerated grid $\mathcal{K}_h^{\mathrm{X}}$ via

$$\tilde{\psi}^{\mathrm{tar}} = \tilde{\psi}^{\mathrm{agg}}, \qquad \tilde{\psi}^{\mathrm{src}} = \mathbf{Q}\tilde{\psi}^{\mathrm{agg}}. \tag{3.78}$$

- *During the simulation run:* On the agglomerated grid $\mathcal{K}_h^{\mathrm{X,agg}}$, we use the cell-local spatial operator vectors in the notation $\mathbf{Op}^{\mathrm{src}} = \mathbf{Op}_j^{\mathrm{src}}$ and $\mathbf{Op}^{\mathrm{tar}} = \mathbf{Op}_j^{\mathrm{tar}}$, see Equation (3.20), for the calculation of the agglomerated cell-local operator

$$\mathbf{Op}^{\mathrm{agg}} = \mathbf{Op}^{\mathrm{tar}} + \mathbf{Q}^{\mathsf{T}}\mathbf{Op}^{\mathrm{src}}. \tag{3.79}$$

We update the solution on the non-agglomerated grid $\mathcal{K}_h^{\mathrm{X}}$ via an explicit time-integration scheme, see Chapter 4, with

$$\tilde{\psi}^{\mathrm{tar}} \leftarrow \tilde{\psi}^{\mathrm{tar}} - \Delta t (\mathbf{M}^{\mathrm{agg}})^{-1}\mathbf{Op}^{\mathrm{agg}}, \tag{3.80}$$

$$\tilde{\psi}^{\mathrm{src}} \leftarrow \mathbf{Q}\tilde{\psi}^{\mathrm{tar}}. \tag{3.81}$$

In summary, this version of the presented cell-agglomeration technique exploits the locality of the outlined operations and allows to reuse the existing operators and data structures. It requires only two additional, cell-local matrix-vector products, see Equations (3.79) and (3.81), in each time-step for fixed interface positions.

3.3.3 Numerical Integration on Cut-Cells

The implementation of an XDG method requires accurate and robust numerical integration techniques for a sub-cell accurate representation of the solution and of the interface itself. Volume and surface integrals on background cells K_j with the boundary ∂K_j, in particular on cut-cells $K_{j,\mathfrak{s}} = K_j \cap \mathfrak{s}$ with the boundary $\partial K_{j,\mathfrak{s}} = \partial(K_j \cap \mathfrak{s})$ and at the interface $\mathfrak{J}$

$$\int_{K_{j,\mathfrak{s}}} (\cdot)\,\mathrm{d}V, \qquad \oint_{\partial K_{j,\mathfrak{s}}} (\cdot)\,\mathrm{d}S, \qquad \text{and} \qquad \oint_{\mathfrak{J}} (\cdot)\,\mathrm{d}S, \tag{3.82}$$

must be evaluated.

The *BoSSS* framework features two different approaches: first, the HMF scheme, originally proposed by Müller et al. (2013) and later extended for the usage in compressible flows with immersed boundaries by Müller et al. (2017); second, an integration technique proposed by Saye (2015) tailored to the *BoSSS* framework by Beck (2018). The HMF scheme can be applied to multiple types of computational cells, for example, triangles, quadrilaterals, tetrahedrons, and hexahedrons, whereas Saye's scheme is restricted to quadrilaterals and hexahedrons. In general, Saye's scheme is faster than the HMF scheme (Kummer et al., 2020).

In the following, we introduce the HMF variant for compressible flow (Müller et al., 2017). The HMF scheme belongs to the class of quadrature schemes, which are built on the solution of the moment-fitting equations (Bremer et al., 2010). The moment-fitting system in a cell K_j is defined as

$$\underbrace{\begin{pmatrix} \phi_{j,1}(\mathbf{x}_1) & \cdots & \phi_{j,1}(\mathbf{x}_L) \\ \vdots & \ddots & \vdots \\ \phi_{j,N}(\mathbf{x}_1) & \cdots & \phi_{j,N}(\mathbf{x}_L) \end{pmatrix}}_{:=A} \underbrace{\begin{pmatrix} \tilde{w}_1 \\ \vdots \\ \tilde{w}_L \end{pmatrix}}_{:=\tilde{w}} = \underbrace{\begin{pmatrix} \int_{K_j} \phi_{j,1}\,\mathrm{d}V \\ \vdots \\ \int_{K_j} \phi_{j,N}\,\mathrm{d}V \end{pmatrix}}_{:=b} \tag{3.83}$$

with the given polynomial basis functions $\phi_{j,n}(\mathbf{x}_l) \in \mathbb{R}^2$ of maximum degree P, a set of nodes $\mathcal{X} = \{\mathbf{x}_l\}_{l=1,\dots,L}$ and a set of the weights $\tilde{\mathcal{W}} = \{\tilde{w}_l\}_{l=1,\dots,L}$ (Bremer et al., 2010). Equation (3.83) is strongly nonlinear in the nodes and linear in the weights (Müller et al., 2017). By solving Equation (3.83) several times while removing insignificant nodes (Xiao and Gimbutas, 2010), a quadrature rule for $\phi_{j,n}(\mathbf{x}_l)$ in the cell K_j is obtained via

$$\int_{(\mathcal{X},\tilde{\mathcal{W}})}^{\mathrm{num}} \phi_{j,n}(\mathbf{x}_l) := \sum_{l=1}^{L} \tilde{w}_l \phi_{j,n}(\mathbf{x}_l). \tag{3.84}$$

For general applications on time-dependent domains, this process is computationally expensive, since the right-hand side b has to be computed in every time-step. Fortunately, it can be

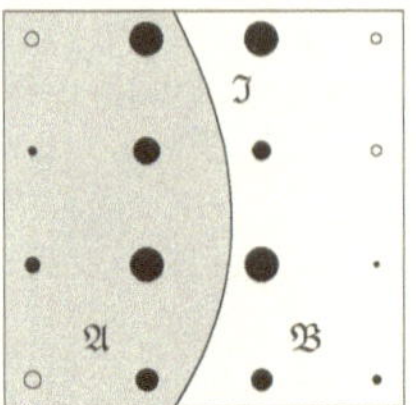
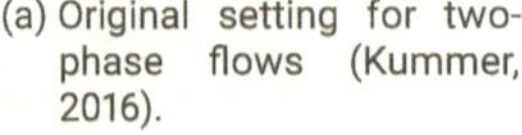
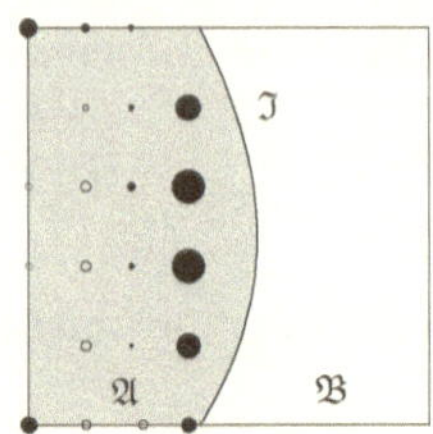

(a) Original setting for two-phase flows (Kummer, 2016).

(b) Modified setting for a DG IBM for compressible flow (Müller et al., 2017).

Figure 3.9: Quadrature nodes and weights of the HMF quadrature scheme for the evaluation of a surface integral in a cut-cell. Filled circles denote positive weights, whereas empty circles denote negative weights. In the DG IBM case, all nodes are placed in the region of interest $\mathfrak{A}$, improving accuracy and avoiding invalid negative state values (adapted from Müller et al., 2017, Figures 2 and 5).

simplified in two ways: a sufficiently accurate and cheap to evaluate approximation of the right-hand side $\mathbf{b}$ in Equation (3.83) is used (*requirement 1*); second, a suitable set of quadrature nodes $\mathcal{X}$ is chosen (*requirement 2*). Thus, Equation (3.83) becomes a linear system

$$\mathbf{A}\tilde{w} = \mathbf{b}, \tag{3.85}$$

which can be over- or under-determined. Of course, the constructed quadrature rule is not optimal in a global sense, but an efficient process can be established leading to sufficiently accurate quadrature rules.

Fulfilling *requirement 1*: The integrals (3.82) are evaluated with a recursive strategy, which is based on the evaluation of a hierarchy of integrals over sub-domains with lower dimensions. This means, that integrals are traced back to their $(D-1)$-lower dimensional cases. For example, a volume integral is traced back to the surface integral and so on. The end of the hierarchy is given by the evaluation of integrals over lines intersected by the interface $\mathfrak{I}$, by finding the roots of the level-set function $\varphi(\mathbf{x})$ on these lines. Müller et al. (2013) define a divergence-free basis, since the interface is only given implicitly and an explicit interface reconstruction would be computationally expensive. The integrands are reformulated using this basis to simplify the integral evaluation. Further details about the evaluation procedure can be found in the original work by Müller et al. (2013).

Fulfilling *requirement 2*: The original HMF variant (Müller et al., 2013) proposes tensor-product Gauss rules on the background cells, as shown in Figure 3.9a. If identical nodes are used on the entire grid, all operations can be vectorized, improving the computational efficiency. Müller et al. (2013) choose the number of Gauss nodes in a way so that Equation (3.85) is under-determined ($L \geq \hat{\beta}N$; L = number of nodes; $\hat{\beta}$ = safety factor; N = number of basis functions). Therefore, the least-squares solution of Equation (3.85) is sufficiently accurate and the safety factor $\hat{\beta}$ increases the robustness and the conditioning of the system. An adapted version with the original node placement strategy was successfully applied in the context of two-phase flows (Kummer, 2016), where an experimental order of convergence (EOC) of $P+1$

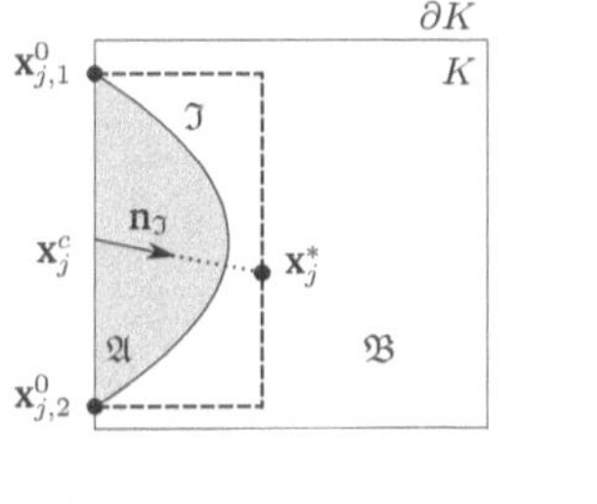

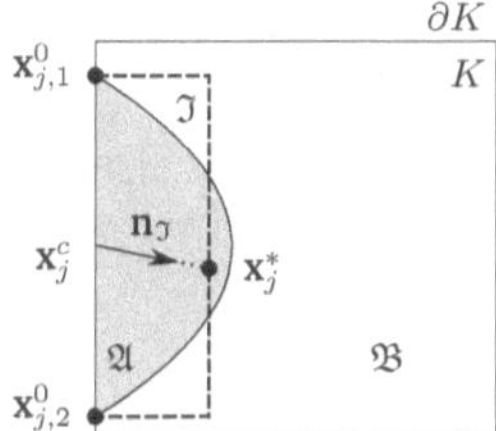

(a) $|\nabla\varphi| > 1$ (steep level-set function).

(b) $|\nabla\varphi| < 1$ (flat level-set function).

Figure 3.10: Bounding boxes for the seeding of quadrature nodes for the HMF quadrature variant for compressible flows (Müller et al., 2017). A bounding box, drawn with a dashed line, is constructed by finding the roots $\mathbf{x}_{j,1}^0$ and $\mathbf{x}_{j,2}^0$ of the level-set function $\varphi(\mathbf{x})$ with the boundary of the background cell ∂K. Additionally, a new point $\mathbf{x}_j^c := 1/2\left(\mathbf{x}_{j,1}^0 + \mathbf{x}_{j,2}^0\right)$ is constructed and moved towards the interface $\mathbf{x}_j^* = \mathbf{x}_j^c + \varphi(\mathbf{x}_j^c)\mathbf{n}_{\mathfrak{J}}$. It is not required that $\varphi(\mathbf{x})$ fulfills the signed distance property. As a result, $\mathbf{x}_j^*$ is not exactly located on $\mathfrak{J}$ (adapted from Müller et al., 2017, Figure 4).

could be reached for a moment-fitting basis of $P_{\mathrm{HMF}} \geq 2P$. Additionally, a modified Cholesky factorization was introduced for the inversion of the mass matrices, since in some cases negative weights lead to a loss of the positive determinacy caused by round-off errors.

Modified placement of quadrature nodes for compressible flow Müller et al. (2017) observed that the node placement in Figure 3.9a cannot be applied in the context of a DG IBM for compressible flow. Placing nodes in the fictitious void region $\mathfrak{B}$ is impractical, since a smooth extension of the solution of the fluid region $\mathfrak{A}$ is needed. This extension is uncontrolled in the void region $\mathfrak{B}$ and, thus, could lead to negative values of the fluid state variables. As a solution, Müller et al. (2017) proposed a new strategy, where the quadrature nodes are solely placed in the fluid region $\mathfrak{A}$, as shown in Figure 3.9b. For that, Müller et al. (2017) developed a simple heuristic strategy where they construct a bounding box around the fluid region $\mathfrak{A}$ for seeding the quadrature points. Figure 3.10 shows the bounding boxes for a steep and a flat level-set function. In a first step, the roots of the level-set function with the background cell boundary $\mathbf{x}_{j,1}^0, \mathbf{x}_{j,2}^0$ are computed. An additional point

$$\mathbf{x}_j^c := \frac{1}{2}\left(\mathbf{x}_{j,1}^0 + \mathbf{x}_{j,2}^0\right) \tag{3.86}$$

is introduced and moved towards the interface

$$\mathbf{x}_j^* = \mathbf{x}_j^c + \varphi(\mathbf{x}_j^c)\mathbf{n}_{\mathfrak{J}}. \tag{3.87}$$

Finally, the points $\mathbf{x}_{j,1}^0$ and $\mathbf{x}_{j,2}^0$ span the bounding box. In this context, $\varphi(\mathbf{x})$ is not enforced to fulfill the signed distance property. As a result, $\mathbf{x}_j^*$ is not located exactly on the interface $\mathfrak{J}$. However, this has a small effect as long as the moment-fitting system (3.83) is still underdetermined when $\hat{\beta}$ is chosen sufficiently large (Müller et al., 2017).

In a next step, equidistant quadrature nodes are seeded inside the bounding box ($L \geq \hat{\beta}N$). All points in the void region $\mathcal{B}$ are removed. By employing this heuristic strategy, Müller et al. (2017) effectively avoided numerically indefinite mass matrices, as reported in Kummer (2016). In this work, we additionally use the variant *Gauss and Stokes Preserving One-Step HMF* (Kummer, 2016), see Appendix A.1.

4 Adaptive Local Time-Stepping for Compressible Flow

The temporal term of time-dependent ordinary differential equations (ODEs) and partial differential equations (PDEs) can be discretized by means of explicit or implicit time-integration schemes. Explicit schemes update the solution solely based on known solutions at previous time-levels, whereas implicit schemes additionally take the unknown solution at the new time-level into account so that an equation system has to be solved. This renders an implicit time-step significantly more computationally expensive than an explicit one. However, this drawback is often negated by the fact that explicit schemes require much more time-steps in total. Furthermore, the CFL-criterion restricts the time-step size of explicit schemes when solving hyperbolic PDEs numerically.

In the context of discontinuous Galerkin (DG) methods, the maximum admissible time-step size of a standard explicit time-stepping scheme is dictated by the smallest grid cell and the order of the polynomial approximation (Cockburn and Shu, 1991; Cockburn and Shu, 2001). For convective terms, the time-step size has a linear dependency on the grid size and the polynomial degree, whereas for diffusive terms, it has a quadratic dependency on the grid size and the polynomial degree as well as a linear dependency on the (artificial) viscosity (Gassner et al., 2008). Thus, a shortcoming of shock-capturing strategies based on artificial viscosity is a severe time-step restriction caused by the second-order diffusive term (Gassner et al., 2008; Gassner, 2009), see Section 5.3. This can decrease the stable time-step size by up to two orders of magnitude, which increases the computational cost enormously.[3]

The severe restriction on the time-step size caused by shock-capturing approaches based on artificial viscosity motivates the use of local time-stepping (LTS) schemes, where the grid cells are grouped into cell clusters according to their local maximum admissible time-step size, while still keeping time accuracy (Dumbser et al., 2007; Sandu and Constantinescu, 2009; Gassner et al., 2011; Grote and Mitkova, 2013). In particular, LTS schemes are well suited in combination with adaptive mesh refinement techniques (Domingues et al., 2008). Besides that, the locality allows for a straightforward combination of the DG method with LTS schemes, for example, as proposed by Kopriva (2009). Based on this work, Winters and Kopriva (2014) developed an explicit LTS scheme for a DG spectral element method on moving grids using an Adams-Bashforth (AB) multi-step method. They confirmed spatial and temporal convergence and provided speed-up and memory estimates for test cases on static and moving grids. Krämer-Eis (2017) extended their approach by implementing a conservative flux interpolation method in the context of compressible flow. Another conservative, second-order LTS formulation was proposed by Krivodonova (2010), who used Heun's method in the context of nonlinear conservation laws.[3]

---

[3]Modified version of Geisenhofer et al. (2019, Section 1).

This chapter deals with a novel variant of an adaptive LTS scheme for the application in a discontinuous Galerkin immersed boundary method (DG IBM) for compressible flow. In Section 4.1, we introduce classical explicit time-integration schemes, such as Runge-Kutta (RK) schemes and AB schemes in Sections 4.1.1 and 4.1.2, respectively, and the equivalent to the CFL-criterion in the context of DG methods. We derive the working principle of a novel adaptive LTS scheme with a variable-coefficient AB scheme for the use on cut-cells in Sections 4.2 and 4.3. We close the chapter with a remark on the computational performance of the adaptive LTS scheme in Section 4.4.

4.1 Explicit Time-Integration Schemes

We start from the semi-discrete system in the form

$$\underbrace{\int_{K_j} \frac{\partial \psi_j}{\partial t} \phi_{j,m} \, \mathrm{d}V}_{\text{temporal term}} - \underbrace{\int_{K_j} \mathbf{f}(\psi_j) \cdot \nabla_h \phi_{j,m} \, \mathrm{d}V + \oint_{\partial K_j} (\mathbf{f}(\psi_j) \cdot \mathbf{n}_\Gamma) \, \phi_{j,m} \, \mathrm{d}S}_{\text{discrete local operator } (\mathbf{Op}_j)_m} = 0 \tag{4.1}$$

with the cell-local approximation $\psi_j = \psi_j(\mathbf{x}, t) = \sum_{n=1}^{N} \psi_{j,n}(t) \, \phi_{j,n}(\mathbf{x})$, see Equations (3.7) and (3.13), respectively. Equation (4.1) has been discretized in space but not yet in time. The approach of discretizing the spatial dimension first is called the *(vertical) method of lines*, in contrast to *Rothe's method* or the *horizontal method of lines*, where the temporal dimension is discretized first. In this work, we limit ourselves to the method of lines. The ODE system (3.16) can be solved by either explicit or implicit time-integration schemes, which we will discuss in the following.

Explicit vs. implicit time-integration schemes Explicit time-integration schemes are suited to solve hyperbolic equations due to their wave-like character. During the temporal update process, the numerical solution in some part of the computational domain is influenced only by the solution in another limited part, see the concept of the domain of dependence/influence in Section 4.1.3. Explicit schemes work well in combination with the locality of the DG method, since the solution known from the previous time-steps has to be exchanged only locally between neighboring cells during the evaluation of the numerical fluxes. In terms of computational costs, explicit schemes are therefore cheap to evaluate, since in contrast to implicit schemes, no (nonlinear) equation system has to be solved. However, explicit schemes lack large stability regions, which results in a prohibitive CFL-restriction. In some cases, implicit time-integration schemes can lead to a speed-up due to their larger stability regions so that significantly less time-steps or only a single time-step has to be performed to compute the solution of a steady problem. In general, implicit schemes offer a better alternative to explicit schemes if the underlying system of ODEs is stiff; for example, this can be caused by strongly nonlinear source terms or algebraic constraints, such as the incompressibility constraint of the incompressible Navier-Stokes equations (Kummer and Müller, 2020). In conclusion, it strongly depends on the application which scheme to choose.

In the following, we limit ourselves to explicit time-integration schemes, since we deal with the hyperbolic Euler equations (2.5) in the context of compressible flow. The derivations in the following sections are based on LeVeque (1999), Müller (2014), Krämer-Eis (2017), and Kummer and Müller (2020).

Multi-step vs. multi-stage methods We recast the temporal term in Equation (3.13) by inserting the cell-local approximation (3.7), which yields

$$\int_{K_j} \frac{\partial \psi_j}{\partial t} \phi_{j,m} \, dV = \int_{K_j} \frac{\partial}{\partial t} \left(\sum_{n=1}^{N} \tilde{\psi}_{j,n}(t) \, \phi_{j,n}(\mathbf{x}) \right) \phi_{j,m} \, dV \tag{4.2}$$

$$= \sum_{n=1}^{N} \frac{\partial \tilde{\psi}_{j,n}}{\partial t} \underbrace{\int_{K_j} \phi_{j,n} \phi_{j,m} \, dV}_{=:(\mathbf{M}_j)_{m,n}} . \tag{4.3}$$

Here, $(\mathbf{M}_j)_{m,n} \in \mathbb{R}^{N,N}$ denotes the symmetric cell-local mass matrix (3.17). Next, we write the semi-discrete form (3.15) in a global, compact form as

$$\frac{\partial \tilde{\psi}_j}{\partial t} + \mathbf{M}_j^{-1} \mathbf{Op}_j(\tilde{\psi}) = 0 , \tag{4.4}$$

which further reduces to

$$\frac{\partial \tilde{\psi}_j}{\partial t} + \mathbf{Op}_j(\tilde{\psi}) = 0 , \tag{4.5}$$

if we assume an orthonormal basis so that $\mathbf{M}_j = \mathbf{I}$. For the derivation of explicit time-integration schemes, we integrate the ODE (4.5) in time, yielding

$$\tilde{\psi}_j(t_{n+1}) = \tilde{\psi}_j(t_n) - \int_{t=t_n}^{t_{n+1}} \mathbf{Op}_j(t, \psi_h(t)) \, dt , \tag{4.6}$$

where $\psi_h(t)$ denotes the discrete solution on the entire computational domain. The integral in Equation (4.6) may be approximated by applying either a multi-step or a multi-stage method. Both have in common that they require operator evaluations solely at time-levels t_n where the solution $\psi_h(t_n)$ is already known or can be interpolated. We briefly discuss the main properties in the following:

- *Multi-step methods:* The simplest and a well-known example of multi-step methods is the *explicit Euler* scheme, where the unknown solution at the new time-level t_{n+1} is calculated by

$$\tilde{\psi}_j(t_{n+1}) = \tilde{\psi}_j(t_n) - \Delta t \, \mathbf{Op}_j(t_n, \psi_h(t_n)) , \tag{4.7}$$

  where $\Delta t = t_{n+1} - t_n$. High-order AB schemes are linear multi-step methods, using the solution history from several previous time-steps. Schemes of this type are computationally cheap to evaluate, but require more memory compared to multi-stage methods, since several previous time-steps have to be saved in the high-order case.

- *Multi-stage methods:* Schemes of this type incorporate the solution at several stages in the time-interval $[t_n, t_{n+1}]$. Compared to multi-step methods, they require less memory, but include more operations due to the evaluation of additional stages. A prominent example are RK schemes, see Section 4.1.1 for details. However, the situation for (diagonally) implicit RK schemes is different, since they require the solution of (nonlinear) equation systems increasing the memory consumption.

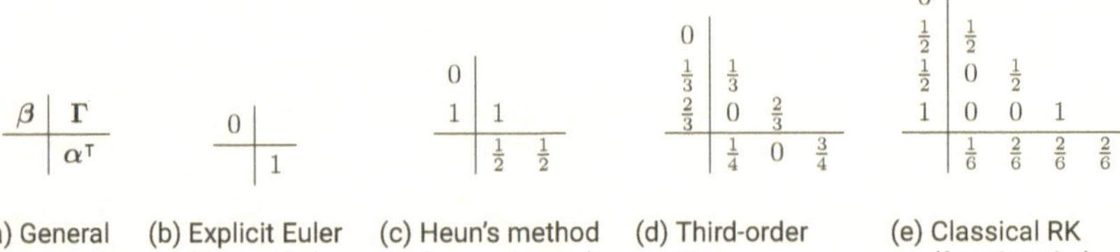

(a) General scheme.	(b) Explicit Euler (first order).	(c) Heun's method (second order).	(d) Third-order TVD.	(e) Classical RK (fourth order).

Figure 4.1: Butcher tableaus for explicit Runge-Kutta (RK) methods. The coefficients $\alpha \in \mathbb{R}^S$, $\beta \in \mathbb{R}^S$, and $\Gamma \in \mathbb{R}^{S,S}$ determine the stability and accuracy of the different variants with S stages (adapted from Müller, 2014, Figure 3.3).

4.1.1 Runge-Kutta Schemes

Explicit RK schemes approximate the solution of the ODE (4.5) by replacing the integral in

$$\tilde{\psi}_j(t_{n+1}) = \tilde{\psi}_j(t_n) - \int_{t=t_n}^{t_{n+1}} \mathbf{Op}_j(t, \psi_h(t))\, \mathrm{d}t \qquad \text{(4.6, repeated)}$$

by a weighted interpolation, which consists of several stages S in the time-interval $[t_n, t_{n+1}]$. Additionally, the stages determine the approximation order of the scheme. In particular, the solution at a new time-level $\tilde{\psi}_j(t_{n+1})$ is calculated staring from a known solution $\tilde{\psi}_j(t_n)$ by

$$\tilde{\psi}_j(t_{n+1}) = \tilde{\psi}_j(t_n) - \Delta t \sum_{s=1}^{S} (\alpha)_s\, \mathbf{k}_s\,, \qquad (4.8\mathrm{a})$$

$$\mathbf{k}_s = \mathbf{Op}_j\left(t_n + (\beta)_s\, \Delta t,\ \tilde{\psi}_j(t_n) + \Delta t \sum_{t=1}^{S} (\Gamma)_{s,t}\, \mathbf{k}_t\right), \qquad (4.8\mathrm{b})$$

$$\Delta t = t_{n+1} - t_n\,. \qquad (4.8\mathrm{c})$$

Variants of RK schemes can be distinguished based on the coefficients $\alpha \in \mathbb{R}^S$, $\beta \in \mathbb{R}^S$, and $\Gamma \in \mathbb{R}^{S,S}$ in Equation (4.8). In particular, these coefficients determine the accuracy and stability of the scheme (Müller, 2014). The operator evaluation (4.8b) at a stage S is denoted by $\mathbf{k}_s$. It is common to state RK schemes in so-called *Butcher tableaus* (Butcher, 1987; Gottlieb and Shu, 1998), which we show up to fourth order in Figure 4.1. We recommend the textbook by Press et al. (2007) for a deeper insight into the topic.

We consider the integral form of the ODE (4.6) where RK schemes can be seen as an integration rule for the integral on the right-hand side in combination with an interpolation rule which allows the evaluation of the integrand on specific nodes (Kummer and Müller, 2020). Precisely, the coefficients α and β denote integration weights and integration nodes, respectively, whereas the matrix Γ describes the interpolation scheme. For purely explicit RK schemes, $\mathbf{k}_s$ only depends on $\mathbf{k}_{\tilde{s}}$ with $\tilde{s} < s$ so that the matrix Γ has to be a strictly lower triangular matrix.

4.1.2 Adams-Bashforth Schemes[4]

AB schemes approximate the integral on the right-hand side in

$$\tilde{\psi}_j(t_{n+1}) = \tilde{\psi}_j(t_n) - \int_{t=t_n}^{t_{n+1}} \mathbf{Op}_j(t, \psi_h(t))\, \mathrm{d}t \qquad \text{(4.6, repeated)}$$

by a polynomial interpolation of the discrete operator at several points in time. Thereby, only known values $\mathbf{Op}_j(\tilde{t}, \psi_h(\tilde{t}))$ with $\tilde{t} < t_{n+1}$ are used. The approximation order Q of an AB scheme depends on the number of the previous time-steps which are used; $Q - 1$ time-steps are needed for an Q-th order AB scheme. The explicit Euler scheme can be seen as an AB scheme of order $Q = 0$.

In the following, we introduce a general AB scheme based on the works by Krämer-Eis (2017) and Geisenhofer et al. (2019). For better readability, we skip the cell index j in Equation (4.6) and additionally use the notations $\tilde{\psi}_n$ and $\mathbf{Op}_n$ instead of $\tilde{\psi}(t_n)$ and $\mathbf{Op}(t_n, \psi_h(t))$, respectively. We replace the integrand in Equation (4.6) by the weighted sum of current and previous operator evaluations $\mathbf{Op}_{n-q}$ with coefficients $\tilde{\beta}_q$, resulting in

$$\tilde{\psi}_{n+1} = \tilde{\psi}_n - \sum_{q=0}^{Q-1} \tilde{\beta}_q\, \mathbf{Op}_{n-q}\,. \qquad (4.9)$$

We determine the coefficients $\tilde{\beta}_q$ by means of Lagrange polynomials

$$\tilde{\beta}_q(t) = \int_{t_n}^{t_{n+1}} \hat{l}_q(s)\, \mathrm{d}s \qquad \text{with} \qquad \hat{l}_q(t) = \prod_{\substack{i=0 \\ i \neq q}}^{Q-1} \frac{t - t_{n-i}}{t_{n-q} - t_{n-i}}\,. \qquad (4.10)$$

As an example, we expand the weighted sum in Equation (4.9) for an order of $Q = 2$, yielding

$$\sum_{q=0}^{1} \tilde{\beta}_q(t)\, \mathbf{Op}_{n-q} = \int_{t_n}^{t_{n+1}} \frac{s - t_{n-1}}{\Delta t_n}\, \mathrm{d}s\, \mathbf{Op}_n + \int_{t_n}^{t_{n+1}} \frac{s - t_n}{-\Delta t_n}\, \mathrm{d}s\, \mathbf{Op}_{n-1}\,, \qquad (4.11)$$

with $\Delta t_n = t_n - t_{n-1}$. We skip the solution of Equation (4.11) and state the final coefficients as

$$\tilde{\beta}_0 = \frac{\Delta t_{n+1}}{2}\left(\frac{\Delta t_{n+1}}{\Delta t_n} + 2\right)\,, \qquad \tilde{\beta}_1 = -\frac{\Delta t_{n+1}}{2}\left(\frac{\Delta t_{n+1}}{\Delta t_n}\right)\,. \qquad (4.12)$$

The final form of a second-order AB scheme with variable coefficients reads

$$\tilde{\psi}_{n+1} = \tilde{\psi}_n - \left(\tilde{\beta}_0 \mathbf{Op}_n + \tilde{\beta}_1 \mathbf{Op}_{n-1}\right)\,. \qquad (4.13)$$

For constant time-steps $\Delta t = \Delta t_{n+1} = \Delta t_n$, Equation (4.9) reduces to

$$\tilde{\psi}_{n+1} = \tilde{\psi}_n - \Delta t\left(\frac{3}{2}\mathbf{Op}_n - \frac{1}{2}\mathbf{Op}_{n-1}\right)\,. \qquad (4.14)$$

---

[4]Modified version of Geisenhofer et al. (2019, Section A.1).

We state the coefficients for a third-order AB scheme

$$\tilde{\beta}_0 = \frac{\Delta t_{n+1}}{6}\, \hat{r}_1\, \hat{r}_0\, \hat{\alpha}\, (2\hat{r}_1 + 3)\,, \tag{4.15}$$

$$\tilde{\beta}_1 = -\frac{\Delta t_{n+1}}{6}\, \hat{r}_1\, \hat{r}_0\, \left(\frac{3}{\hat{\alpha}} + 2\hat{r}_1\right)\,, \tag{4.16}$$

$$\tilde{\beta}_2 = \frac{\Delta t_{n+1}}{6}\, \left(2\hat{\alpha}\hat{r}_1^2 + 3\hat{r}_1\, (\hat{\alpha} + 1) + 6\right)\,, \tag{4.17}$$

with the auxiliary variables

$$\hat{r}_0 = \frac{\Delta t_n}{\Delta t_{n-1}}\,, \qquad \hat{r}_1 = \frac{\Delta t_{n+1}}{\Delta t_n}\,, \qquad \hat{\alpha} = \frac{\hat{r}_0}{\hat{r}_0 + 1}\,. \tag{4.18}$$

High-order AB schemes ($Q > 1$) are not self-starting, since no previous time-steps are available. As a solution, we use a RK scheme of at least the same order for the start-up phase in order to not introduce any additional temporal error.

4.1.3 Time-Step Restrictions

The original Courant-Friedrichs-Lewy (CFL)-restriction, traced back to the work of Courant et al. (1928), denotes a necessary condition for stability when applying an explicit Euler time-integration scheme to a class of linear hyperbolic PDEs (LeVeque, 1999). It states that the maximum admissible time-step size Δt depends on the largest information propagation speed $|\mathbf{u}|$ and on a characteristic length scale h, which is usually given by the computational grid, resulting in the expression

$$\Delta t \leq C_{\mathrm{CFL}}\frac{h}{|\mathbf{u}|}\,, \tag{4.19}$$

where the constant $0 < C_{\mathrm{CFL}} \leq 1$ depends on the spatial discretization. Roughly speaking, Equation (4.19) states that the information is not allowed to travel farther than one grid cell in one time-step.

Domain of dependence and domain of influence We consider the one-dimensional linear advection equation

$$\frac{\partial \psi(x_1, t)}{\partial t} + u_1\frac{\partial \psi(x_1, t)}{\partial x_1} = 0\,, \tag{4.20}$$

where $u_1 = \text{const.}$, as an example of a one-dimensional first-order linear PDE. We introduce a characteristic curve $x_{\mathrm{c}}(t)$ in the x_1, t-plane along which the solution $\psi(x_1, t)$ is constant and depends solely on the initial condition $x_c^0 = x_{\mathrm{c}}(t_0) = \psi(x_0, t_0)$. Thus, the property

$$\frac{\mathrm{D}\psi(x_{\mathrm{c}}(t), t)}{\mathrm{D}t} = \frac{\partial \psi}{\partial t} + \frac{\mathrm{d}x_{\mathrm{c}}(t)}{\mathrm{d}t}\frac{\partial \psi}{\partial x_{\mathrm{c}}} = 0 \tag{4.21}$$

holds. Comparing Equation (4.21) to Equation (4.20) yields the expression

$$u_1 = \frac{\mathrm{d}x_{\mathrm{c}}(t)}{\mathrm{d}t} \tag{4.22}$$

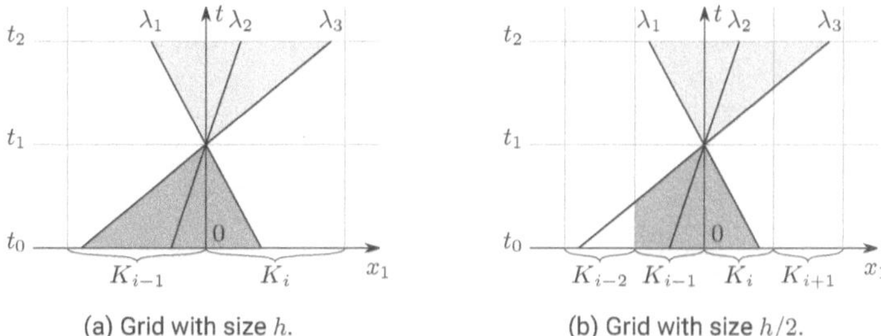

(a) Grid with size h.   (b) Grid with size $h/2$.

Figure 4.2: Domain of dependence (dark gray) and domain of influence (light gray) on two different grids. We consider a system of three first-order linear one-dimensional PDEs with their characteristic speeds λ_i. (a) The solution at $(0, t_1)$, which is the intersection point of all characteristics, is influenced by the initial conditions in the respective cells. (b) The CFL-restriction (4.19) is violated, since the solution at $(0, t_1)$ is not influenced by the initial condition in K_{i-2} (adapted from Kummer and Müller, 2020, Section 4.3).

so that the PDE (4.20) reduces to the ODE (4.22) along a characteristic curve. For the presented case of $u_1 = \text{const.}$, the ODE (4.22) can be solved by a simple integration using the initial condition $x_c^0 = x_c(t_0)$, resulting in the characteristic curve

$$x_c(t) = x_c^0 + u_1 t \, , \tag{4.23}$$

where u_1 denotes the characteristic speed.

For illustration purposes, we now consider the characteristics of a system of three one-dimensional first-order linear PDEs with the corresponding characteristic speeds λ_1, λ_2, and λ_3. Figure 4.2 shows an x_1, t-diagram on two different grids with the characteristic curves at three time-levels. If we look at the state at $(0, t_1)$, which is the intersection point of all characteristics in Figure 4.2a, we recognize that the solution is influenced by the dark gray region (*domain of dependence*) and will influence the light gray region (*domain of influence*). In Figure 4.2b, we clearly see that the CFL-restriction is violated, since the solution at $(0, t_1)$ is not influenced by the initial condition in cell K_{i-2}. As a result, we cannot expect the numerical solution to converge to the true solution of the underlying PDE with the given initial conditions.

Time-step restrictions in the context of discontinuous Galerkin methods[5] In the context of a DG method, the maximum admissible time-step size is dictated by the characteristic length scale h and the order of the approximating polynomial P (Cockburn and Shu, 1991). Cockburn and Shu (2001) defined a DG equivalent to the original CFL-restriction (4.19) for convection dominated problems as

$$\Delta t_c \le \frac{C_{\text{CFL}}}{2P + 1} \frac{h}{|\mathbf{u}| + a} \, , \tag{4.24}$$

where $0 < C_{\text{CFL}} \le 1$ is a problem-dependent constant and $|\mathbf{u}| + a$ describes the fastest propagation velocity of the hyperbolic Euler equations (2.5).

---

[5]Modified version of Geisenhofer et al. (2019, Section 5.1).

_____

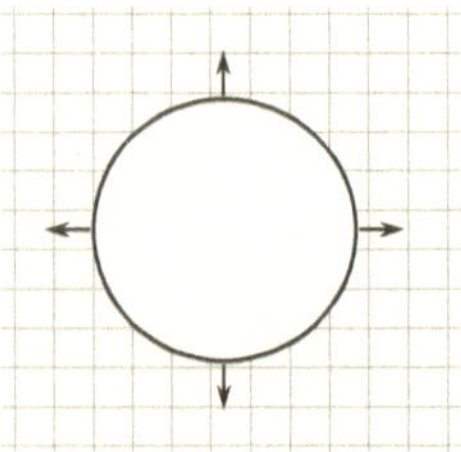

(a) The cylinder is slightly shifted to produce different cut configurations.

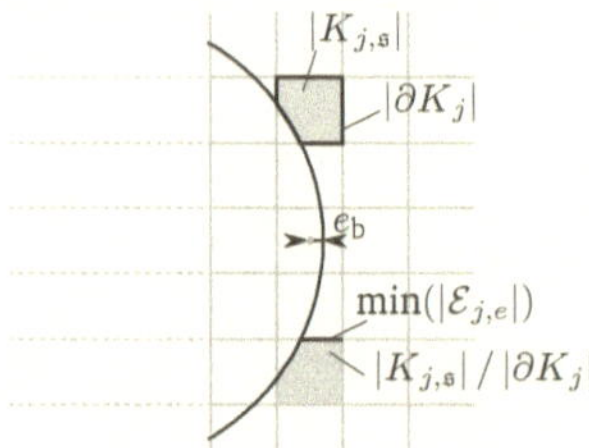

(b) Three different measures for the characteristic length scale h.

Figure 4.3: Numerical experiment to determine the cell-local characteristic length scale h on cut-cells (Krämer-Eis, 2017). A steady flow around a cylinder is investigated, where the maximum stable time-step size is determined empirically and afterwards compared to the theory for different measures of h (adapted from Geisenhofer et al., 2019, Figure 3).

For diffusion dominated problems, Gassner et al. (2008) derived a similar expression

$$\Delta t_d \leq \frac{C_{\mathrm{DFL}}}{(2P+1)^2} \frac{h^2}{\sqrt{D}\,\mu\,\max\left(\frac{4}{3},\frac{\gamma}{\mathrm{Pr}}\right)}, \tag{4.25}$$

where $0 < C_{\mathrm{DFL}} \leq 1$ is another scaling factor, D is the number of spatial dimensions and $\mathrm{Pr} = \mu_\infty c_p/k_\infty$ is the Prandtl number. Appropriate suggestions for the scaling factors C_{CFL} and C_{DFL} can be found in their work (Gassner et al., 2008).

For convection-diffusion problems, we use the harmonic sum

$$\Delta t = \frac{1}{\frac{1}{\Delta t_c} + \frac{1}{\Delta t_d}}. \tag{4.26}$$

Equation (4.26) is a more reliable measure than the minimum of the pure convective time-step restriction (4.24) and the pure diffusive time-step restriction (4.25) (Watkins et al., 2016).

In the past, various authors (Gassner et al., 2008; Kubatko et al., 2008; Toulorge and Desmet, 2011; Watkins et al., 2016) widely studied the calculation of the maximum stable time-step size on structured and unstructured boundary-fitted grids. By contrast, this topic was hardly addressed for a cut-cell grid in the context of a DG IBM, since it is challenging to determine a suitable choice of h on cut-cells. Krämer-Eis (2017) investigated this issue by comparing three different cell measures

(i) $|K_{j,\mathfrak{s}}|/|\partial K_j|$ (fluid volume fraction),

(ii) $2e_\mathrm{b}$ (e_b is the distance from the barycenter of a cell to the closest edge or interface),

(iii) $\min(|\mathcal{E}_{j,e}|)\,(|K_{j,\mathfrak{s}}|/|\partial K_j|)$ (shortest edge × fluid volume fraction),

for a steady compressible flow around a cylinder, where the cylinder is embedded in a uniform grid, see Figure 4.3. In order to quantify the influence of the different measures for different cut configurations, the cylinder was slightly shifted in horizontal and vertical direction. For all cut scenarios, the maximum stable time-step sizes were empirically determined by numerical simulations. The theoretical time-step size was additionally calculated using Equations (4.24) and (4.25). The average percentage deviations from the empirically determined time-step size are $13.6\,\%$ (measure (i)), $33.4\,\%$ (measure (ii)) and $25.6\,\%$ (measure (iii)). This indicates that measure (i) is a suitable candidate for a DG IBM, while measures (ii) and (iii) underestimate the maximum stable time-step sizes in cut-cells and, thus, lead to an unnecessarily large number of time-steps.

We abstain from going into details even further as the presented choice of h is sufficient for the purpose of this work. Krivodonova and Qin (2013) showed that there is no generally valid choice of h for all cut scenarios. Furthermore, they state that the number and in particular the position of small cells on a coarse grid with otherwise identical cells significantly influence the global maximum stable time-step size.

For the remainder of this work, we choose the cell-local characteristic length scale of a cell K_j as follows

$$
h_j := \begin{cases} \min\left(|\mathcal{E}_{j,e}|\right), & \text{if } K_j \text{ is a standard or agglomeration target cell}, \\ |K_{j,\mathfrak{s}}| \, / \, |\partial K_j|, & \text{if } K_j \text{ is a non-agglomerated cut-cell}. \end{cases}
$$

$$(3.57, \text{repeated})$$

4.2 Derivation of the Basic Local Time-Stepping Scheme[6]

In a global time-stepping method, the cell with the smallest time-step size dictates the global time-step size. For large differences in the local time-step sizes, the computational costs grow enormously, for example, as in the case of a cut-cell DG IBM with a two-step shock-capturing strategy based on artificial viscosity, see Chapter 5. Here, the time-step size in non-agglomerated cut-cells which also contain a shock may differ by up to two orders of magnitude from the time-step size in standard background cells in the bulk flow. This motivates the application of an adaptive LTS scheme where the clustering is based on the local time-step sizes and is dynamically adapted to the physical changes in unsteady flows, see Section 4.3. As a basis, we introduce a generic LTS scheme in this section.

LTS schemes belong to the class of multi-rate methods which solve ODEs on different time-scales. Our implementation of an explicit LTS scheme is based on the approach by Winters and Kopriva (2014). Their scheme is a simplification of the multi-rate linear multi-step method of Gear and Wells (1984), which was combined with two-step AB schemes, for example, by Stock (2009). We use a variable-coefficient AB scheme to evolve the solution in time, see Section 4.1.2. For the standard static case with constant coefficients, Krämer-Eis (2017) developed a flux correction procedure which resulted in a conservative LTS formulation.

---

[6]Modified version of Geisenhofer et al. (2019, Section 5.2).

Assumptions   We make the following assumptions based on the works by Winters and Kopriva (2014) and Krämer-Eis (2017):

- The grid cells are grouped into a *set of cell clusters* $\mathcal{M} = \{M_1, \dots, M_M\}$, where $M_i \subseteq \mathcal{K}_h^{\mathrm{X,agg}}$ or $M_i \subseteq \mathcal{K}_h$. The clustering is based on a cell-local characteristic measure. In this work, we use the maximum admissible cell-local time-step size Δt_j.

- The time-levels of the cell clusters are synchronized at *synchronization levels*.

- The time-scales of the cell clusters differ only by *integer factors*.

- The notation K_j covers uncut background cells and cut-cells $K_{j,\mathfrak{s}}$ for a better readability in this section.

Derivation   Subsequently, we briefly introduce a multi-rate time-integration method for a coupled ODE system $\partial \mathbf{y} / \partial t = \mathbf{F}(t, \mathbf{y})$, assuming a fast and a slow time-scale with superscripts $(\cdot)^{\mathrm{f}}$ and $(\cdot)^{\mathrm{s}}$, respectively (Gear and Wells, 1984; Stock, 2009; Winters and Kopriva, 2014). Thus, the state vector and the right-hand side can be divided into *slow*, *fast*, and *coupled* components

$$\mathbf{y} = \begin{bmatrix} \mathbf{y}^{\mathrm{f}} \\ \mathbf{y}^{\mathrm{s}} \end{bmatrix}, \tag{4.27}$$

and

$$\mathbf{F} = \begin{bmatrix} F_{\mathrm{ff}} & F_{\mathrm{fs}} \\ F_{\mathrm{sf}} & F_{\mathrm{ss}} \end{bmatrix}, \tag{4.28}$$

where $\mathbf{y}^{\mathrm{f}}$ has to be updated using smaller time-steps than $\mathbf{y}^{\mathrm{s}}$ so that

$$\Delta t_{\mathrm{f}} < \Delta t_{\mathrm{s}}. \tag{4.29}$$

In Equation (4.28), the main diagonal components F_{ff} and F_{ss} can be calculated immediately, as they are on the same time-scales (Winters and Kopriva, 2014). In the DG method, this step corresponds to the evaluation of volume and surface integrals. However, the coupled terms F_{sf} and F_{fs} require an inter-/extrapolation procedure from slow to fast time-scales and vice versa. In particular, this includes the evaluation of surface integrals between neighboring cells. An interpolation is needed in order to calculate the coupled component F_{sf}, whereas an extrapolation is needed to calculate F_{fs}, which is an inherently unreliable calculation (Winters and Kopriva, 2014). The extrapolation is needed since the time-levels of the fast scales are usually smaller than those of the slow scales during the evaluation process. Due to the weak coupling between cells in the DG method, the extrapolation F_{fs} can be treated separately, see Winters and Kopriva (2014) for details. Consequently, Equation (4.28) reduces to

$$\mathbf{F} = \begin{bmatrix} F_{\mathrm{ff}} & 0 \\ F_{\mathrm{sf}} & F_{\mathrm{ss}} \end{bmatrix}. \tag{4.30}$$

In the end, we only need to interpolate the values from slow to fast time-scales F_{sf} at intermediate time-levels. In the case of a DG method, this step corresponds to the evaluation of the numerical fluxes. In total, fewer operator evaluations are needed for an LTS scheme than for a standard global AB scheme, whose performance is limited by the cell with the smallest time-step size. However, additional work is required in order to calculate the time-interpolants and the coupled terms.

We use the harmonic sum (4.26) to determine the cell-local maximum admissible time-step $\Delta \tilde{t}_j$ for a cell on the standard grid ($K_j \in \mathcal{K}_h$) or on the agglomerated grid ($K_{j,\mathfrak{s}} \in \mathcal{K}_h^{\mathrm{X,agg}}$). Next, we cluster the cells into sets of similar time-step sizes $\mathcal{M} = \{M_1, \ldots, M_M\}$ by using the K-means algorithm (Macqueen, 1967). Then, these clusters fulfill the property

$$\underbrace{\min_{K_j \in M_1} \Delta \tilde{t}_j}_{:=\Delta t_1} > \min_{K_j \in M_2} \Delta \tilde{t}_j > \ldots > \min_{K_j \in M_M} \Delta \tilde{t}_j , \tag{4.31}$$

where Δt_1 is the maximum admissible time-step size of the slowest cluster M_1, which is the one with the largest time-step size. Additionally, we assume that the time-scales Δt_m between different clusters differ only by integer factors f_m

$$\Delta t_m = \frac{\Delta t_1}{f_m} , \qquad m = 2, \ldots, M , \tag{4.32}$$

with

$$f_m = \mathrm{ceil} \left\lceil \frac{\Delta t_1}{\min\limits_{K_j \in M_m} \Delta \tilde{t}_j} \right\rceil , \qquad f_m \in \mathbb{N} . \tag{4.33}$$

Equation (4.32) results in the beneficial fact that all clusters are synchronized at synchronization levels depending on the integer factor of the smallest cluster f_M, yielding

$$t^{n+1} = t^n + \Delta t_1 , \qquad n \in \mathbb{N} . \tag{4.34}$$

Flux interpolation   In general, if we update the solution in every cluster M_m with its respective time-step size Δt_m after all clusters have reached a synchronization level, all solutions will be on different time-levels, as shown in Figure 4.4a. In order to update the solution in M_3, we have to interpolate the solution of the outer values ψ_h^{out} at the intermediate time t^*, as shown in Figure 4.4b. For this, we use the time history of the solutions in M_1, which are directly available in an AB method. Next, we can use the interpolated value of ψ_h^{out} in order to evaluate the numerical flux $\hat{F}\left(\psi_h^{\mathrm{in}}, \psi_h^{\mathrm{out}}, \mathbf{n}_\Gamma\right)$ across the boundary of the cells/clusters $M_1 \leftrightarrow M_3$, as shown in Figure 4.4c. Aligning well with a DG method, the LTS scheme only needs the interpolant of the direct neighbor without creating any additional larger stencils or buffers. Due to the interpolation procedure, the LTS scheme is non-conservative. Krämer-Eis (2017) developed a flux correction procedure, regaining a conservative formulation based on an adaption of the numerical fluxes.

Start-up phase   At the initial synchronization level t^0, AB methods for an explicit order of $Q > 1$ are not self-starting, since no time history is available yet, see Section 4.1.2. As a solution, we use a global explicit Runge-Kutta scheme of one order higher than the AB scheme for the start-up phase in order to guarantee that the numerical error of the RK scheme is small compared to the one introduced by the LTS scheme (Winters and Kopriva, 2014). As soon as the necessary time history is available, we switch back to the AB scheme.

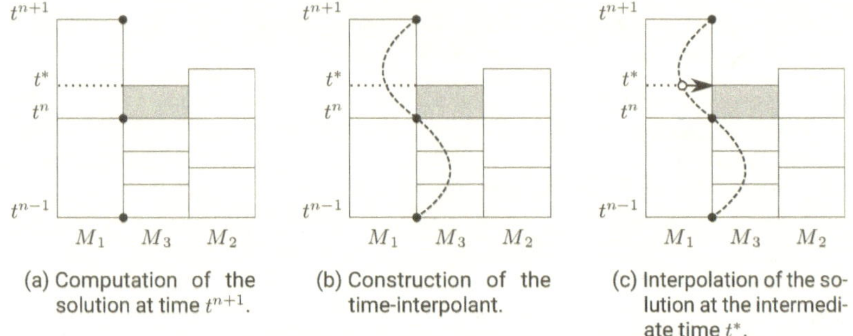

(a) Computation of the solution at time t^{n+1}.　　(b) Construction of the time-interpolant.　　(c) Interpolation of the solution at the intermediate time t^*.

Figure 4.4: Flux interpolation of the LTS scheme. We consider a clustering $\mathcal{M} = \{M_1, M_2, M_3\}$ consisting of single one-dimensional cells with the time-step sizes $\Delta t_2 = \Delta t_1/2$ and $\Delta t_3 = \Delta t_1/3$. The solution history of cluster M_1 is marked with filled circles, the time-interpolant is drawn as a dashed line, and the interpolated intermediate solution at time t^* is marked with an empty circle (adapted from Krämer-Eis, 2017, Figure 3.5).

4.3 Adaptive Reclustering[7]

In unsteady supersonic flows, shocks and other discontinuous phenomena propagate through the computational domain. Accordingly, the grid cells, where artificial viscosity is added, and, thus, the cell with small time-step sizes change in time. This makes an adaption of the LTS clustering necessary. A static clustering would negate the basic idea of an LTS scheme, namely a reasonable update of the cells according to their local time-step sizes. As a solution, we introduce a reclustering interval $I_{\mathrm{LTS}} \in \mathbb{N}$, at which we rebuild the clustering and copy the time history of all cells when using an AB scheme of order $Q > 1$. The history contains cell-local information about the update times and time-step sizes, the change rates, which is the weighted sum in Equation (4.9), and the DG coefficients from previous time-steps. An LTS method relies on an *evolve condition*, which determines if a cell can be evolved in time. We use the evolve condition by Winters and Kopriva (2014), who define an intermediate time t^* between two synchronization levels t^n and t^{n+1}, $t^* \in [t^n, t^{n+1}]$: *"If the local time on cell K_j is equal to t^*, then the cell is ready to evolve one local time-step"*.

In Figure 4.5, we outline the procedure of the adaptive LTS scheme for the two time-intervals $t^n \to t^{n+1}$ and $t^{n+1} \to t^{n+2}$. We set $I_{\mathrm{LTS}} = 1$ and assume a one-dimensional grid with four equidistant cells for the sake of simplicity. Note that in the x_1, t-diagram, the height of the cells refers to the time-step size in the specific cluster according to Equations (4.32) and (4.33). Cells with a small height can be interpreted as cells which contain a shock and, thus, artificial viscosity, and/or as non-agglomerated cut-cells.

We present a pseudocode for a generic implementation between two synchronization levels $t^n \to t^{n+1}$ in Algorithm 4.1. During the start-up phase, we employ a RK scheme to generate the necessary solution history for AB schemes of order $Q > 1$, see Section 4.2. Additionally, we create the data structures for storing the histories of, for example, the cell-local update times or the DG coefficients, in order to execute the adaptive LTS scheme. The adaptive reclustering in

---

[7]Modified version of Geisenhofer et al. (2019, Section 5.3).

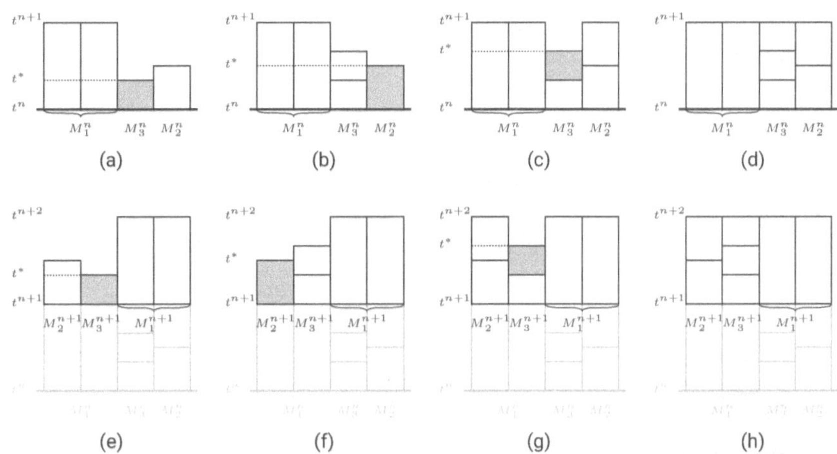

Figure 4.5: Illustration of the adaptive LTS procedure in an x_1, t-diagram. We consider two time-intervals $[t^n, t^{n+1}]$ (first row) and $[t^{n+1}, t^{n+2}]$ (second row) on a one-dimensional grid with four cells. The cell clustering is denoted by $\mathcal{M} = \{M_1, M_2, M_3\}$. The height of a x_1/t_m-cell corresponds to the local time-step size. (a) - (c) and (e) - (g) Filled gray cells are at intermediate time t^* and will be updated next. (d) and (h) All cells are at a synchronization level. (e) Additionally: A reclustering is performed based on the change of the time-step restrictions (adapted from Geisenhofer et al., 2019, Figure 4).

Algorithm 4.1: Pseudocode of the adaptive LTS scheme for the integration between two synchronization levels $t^n \rightarrow t^{n+1}$. The adaptive reclustering is controlled by a user-specified interval I_{LTS} and only executed if it is induced by the physical changes in the flow. The use of histories is only required for Adams-Bashforth schemes of order $Q > 1$ (adapted from Geisenhofer et al., 2019, Algorithm 1).

```
 1: procedure DoALTSStep( )
 2:     UpdateTimeStepSizes()                          ▷ Computed in all fluid cells, Equation (4.26)
 3:
 4:     if startUpPhase then
 5:         UseRungeKuttaTimeStepper()                                          ▷ If order Q > 1
 6:         CreateHistories()                  ▷ Cell-local update times, change rates, DG coefficients
 7:     else if isReclusteringInterval and reclusteringIsNecessary    ▷ Larger changes in the flow field?
 8:         CreateNewClustering()
 9:     end if
10:
11:     UpdateClusterTimeStepSizes()                 ▷ Performed at every synch. level, Equation (4.32)
12:     DoAdamsBashforthTimeIntegration()                              ▷ Evolve all clusters in time
13:     UpdateHistories()
14: end procedure
```

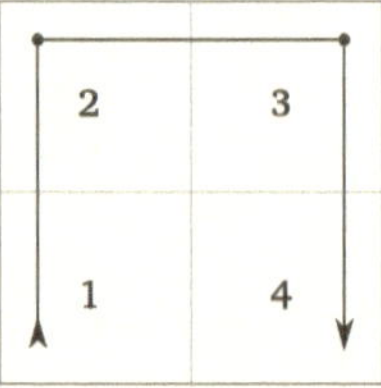

(a) Step 1: subdivision.

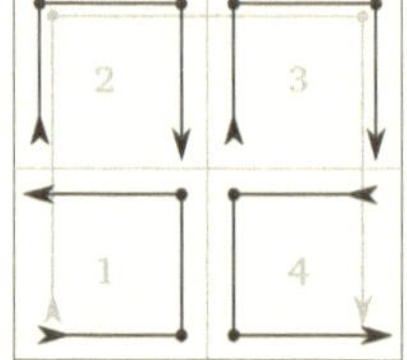

(b) Step 1: transformation.

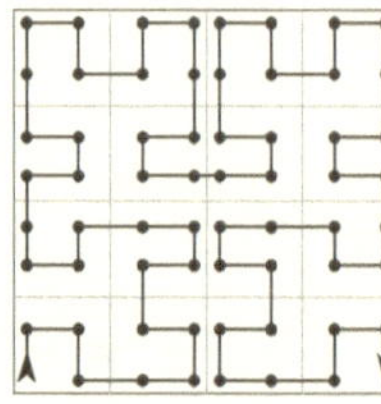

(c) Step 2: subdivison, transformation, and connection.

Figure 4.6: Recursive refinement of a Hilbert curve. (a) The domain is subdivided in 2^D subdomains. (b) In every subdomain, a transformed version of the Hilbert curve is created which can be connected to a coherent curve. (c) This procedure can be recursively repeated up to any desired refinement level (adapted from Weber, 2018, Figure 2.3).

time is performed at a user-specified interval I_{LTS} only if it is induced by the physical changes in the flow in order to save computational costs. In this case, the relevant histories are copied and a new LTS clustering is created. Each cluster is integrated in time with its respective stable time-step size, before the histories are updated.

4.4 Computational Performance

The parallelization of LTS schemes remains a delicate issue to solve. In Section 4.3, we presented an adaptive LTS scheme, where the cells are grouped into cell clusters according to their local time-step size. Figure 4.5 clearly shows that during a simulation run, every cell can have its own individual total number of updates due to the adaptive reclustering after several time-steps. This fact suggests the connection to adaptive mesh refinement techniques, where a local refinement in space is applied (Berger and Oliger, 1984; Rannabauer et al., 2018). Consequently, the presented LTS scheme can be seen as an 'adaptive mesh refinement in time'. From a computational point of view, the LTS scheme increases the computational load locally. As a remedy, grid-partitioning techniques can be applied in order to dynamically distribute the computational load equally among multiple processors. The parallelization in the employed *BoSSS* framework is facilitated by the Message Passing Interface (MPI).

Grid-partitioning by means of Hilbert curves During the course of a Master's book (Weber, 2018), we investigated the influence of grid-partitioning on the computational performance of the presented adaptive LTS scheme in terms of partitioning quality and speed-up. Hereby, we used a grid-partitioning technique based on space-filling Hilbert curves (Butz, 1969; Hindenlang, 2014) and the ParMETIS/METIS graph partitioning framework (Karypis and Kumar, 1998; Karypis and Kumar, 1999; Karypis, 2020), which served as a reference.

A space-filling Hilbert curve traverses a computational domain as shown in Figure 4.6. It has the following properties among others (Haverkort, 2012):

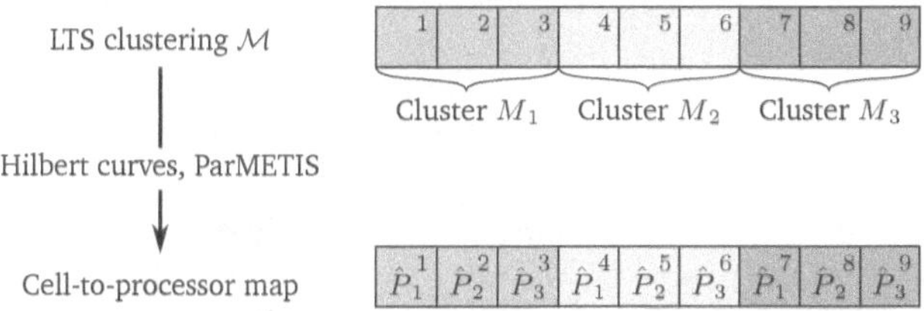

Figure 4.7: Cell-to-processor mapping for the adaptive LTS scheme. We assume a clustering $\mathcal{M} = \{M_1, M_2, M_3\}$ consisting of cells K_j with their global indices $j = 1, \ldots, 9$. The cells of each cluster M_m are distributed equally among the processors $\hat{P}_i$ ($i = 1, 2, 3$) in order to split the computational load evenly.

- A Hilbert curve is fractal. With every recursive step, the Hilbert curve is refined by a factor of two in every spatial dimension.

- A Hilbert curve is self-avoiding. A point on the computational domain $\mathbf{x} \in \Omega$ is visited only once by the Hilbert curve.

- A Hilbert curve is order-preserving. The order of the points along the Hilbert curve does not change with an increased refinement level.

In the first step shown in Figure 4.6a, the grid is subdivided into four equal subdomains. On each subdomain, a transformed version of the base pattern of the Hilbert curve is created by means of rotation, scaling, and mirroring operations, as shown in Figure 4.6b. This process can be recursively repeated until the desired refinement is reached, see Figure 4.6c. The mapping of a cell to the Hilbert curve is realized by using an auxiliary grid with a total resolution of a 64-bit-integer. This results in a grid size of $h = 2^{64/D}$ for D spatial dimensions. We consider an exemplary cell with its barycenter located at $(20.7, 5.4)$. The coordinates of the barycenter are rounded and assigned to the point $(20, 5)$ on the auxiliary grid. The actual mapping to a Hilbert index is performed by means of the Butz-Moore algorithm (Butz, 1969). For the scope of this work, it is sufficient to consider the Hilbert index as a numbering index in the traversing direction of the Hilbert curve. The order in which cells are distributed to the different processors is uniquely defined by the marching direction of the Hilbert curve. This is beneficial for the grid-partitioning quality, in particular for a low inter-processor communication, since the next cell on the Hilbert curve is always a neighboring cell.

In the context of the applied adaptive LTS scheme, we distribute the cells to the different processors in order to split the computational cost evenly, see Figure 4.7. Note that the order of the processors within a cluster in the *Cell-to-processor map* may vary for different partitioning approaches. The number of updates which a cell experiences during the run-time can be used as a measure for the computational costs. As an example, we assume a clustering of three clusters $\mathcal{M} = \{M_1, M_2, M_3\}$ with the time-step sizes Δt_1, $\Delta t_2 = \Delta t_1/2$, and $\Delta t_3 = \Delta t_1/3$ so that $\Delta t_3 < \Delta t_2 < \Delta t_1$. If all cells in M_1 are updated once, all cells in M_2 have to be updated twice, and all cells in M_3 have to be updated three times. This makes the cells in M_3 three times, and the cells in M_2 twice as expensive than the cells in M_1 from a computational point of view, when assuming constant costs for any cell evaluation. Thus, one approach to split the computational load evenly is to distribute the cells of each cluster in the clustering $\mathcal{M}$ equally among the processors.

Grid-partitioning quality We show a simulation of a double Mach reflection (DMR), see Section 5.5.4, with a polynomial degree of $P = 0$ on a grid with 800×200 cells. The adaptive LTS scheme is activated for a reclustering interval of $I_{\mathrm{LTS}} = 10$ with an order of $Q = 1$ and a clustering consisting of three clusters ($C_{\mathrm{init}} = 3$). Figure 4.8 shows the results after 300 time-steps. In general, the Hilbert algorithm distributes the cells along the traversing direction of the Hilbert curves to the different processors without interruption. It proceeds to the next processor if the maximum number of cells on the current processor is reached. In the underlying test case, the auxiliary constraint presented in Figure 4.7 has to be additionally considered. It forces each processor to have (approximately) the same number of cells of each LTS cell cluster. This leads to a frayed grid-partitioning in the area $[1.8, 2.2] \times [0, 0.4]$, where the cells belong to different LTS cell clusters. Based on the refinement level, the Hilbert curve may leave or enter regions with cells of the same LTS cell cluster several times, see also Figure 4.6c. The ParMETIS/METIS algorithm yields a more uniform distribution in this case. Note that a grid-partitioning which is best in terms of inter-processor communication features as few border cells between different processors as possible. This holds true for ParMETIS/METIS in this case. Further improvements for the grid-partitioning based on Hilbert curves may be a repair method linking unconnected parts or a different approach assigning the grid cells to the Hilbert curve. The reader is referred to the work by Weber (2018) for more details.

Speed-up The speed-up of parallelized computational simulations can be measured by means of Amdahl's law (Amdahl, 1967). We follow the derivation given by Weber (2018). In general, the total run-time of a computational algorithm T, being executed on multiple processors $\hat{P}$, can be divided into a sequential part t_S and a parallel part t_P

$$T = t_S + t_P. \tag{4.35}$$

We expect that the total run-time on several processors $T(\hat{P})$ decreases if we increase the number of processors $\hat{P}$ when executing a program for a fixed problem size. The run-time needed for the execution of the sequential part on one processor $T(\hat{P} = 1)$ is a lower bound for the total run-time if we neglect the communication overhead between different processors. We define the speed-up for a fixed problem size as the ratio of the sequential run-time divided by the parallel run-time

$$\hat{S} = \frac{T(1)}{T(\hat{P})} = \frac{1}{t_S + \frac{t_P}{\hat{P}}} = \frac{1}{t_S + \frac{1-t_S}{\hat{P}}}, \tag{4.36}$$

where the parallel run-time can be substituted with $t_P = 1 - t_S$, assuming a total run-time of $T = 1$. Figure 4.9 shows the speed-up (4.36) for several sequential parts $t_S = \{0.1, 0.05, 0.01, 0.001\}$. The speed-up is theoretically limited by $\lim_{\hat{P} \to \infty} \hat{S}(\hat{P}) = 1/t_S$. We define the average speed-up as

$$\hat{S}_{\mathrm{avg}}(\hat{P}) = \frac{\overline{T}(1)}{\overline{T}(\hat{P})}, \tag{4.37}$$

calculated by the arithmetic means of the run-times $\overline{T}(\hat{P})$.

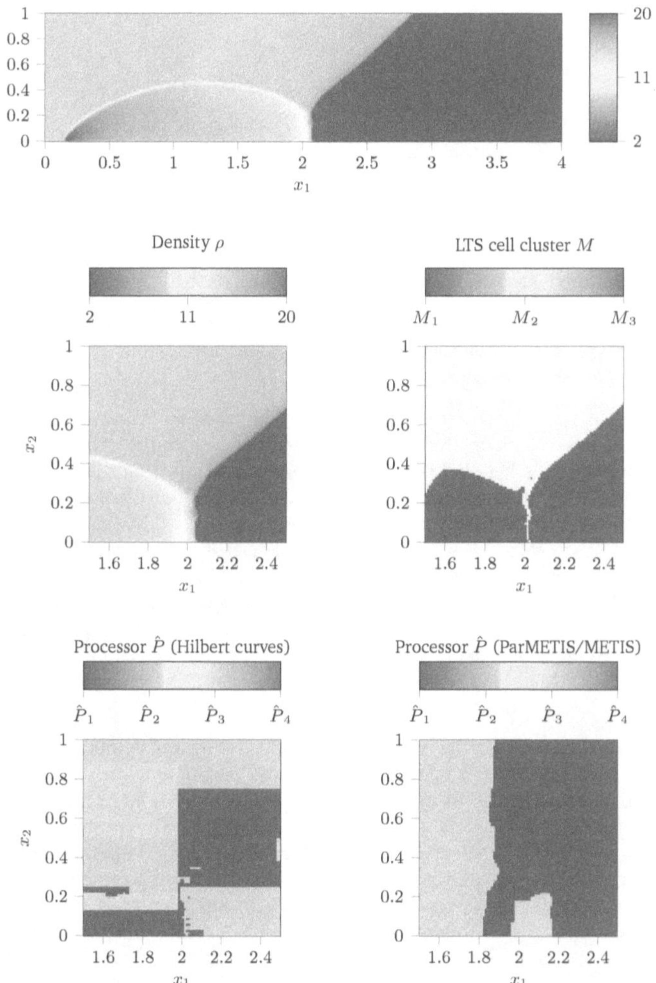

Figure 4.8: Grid-partitioning for the adaptive LTS scheme: space-filling Hilbert curves (Butz, 1969) vs. the ParMETIS/METIS framework (Karypis, 2020). The Hilbert algorithm tries to distribute the cells to the different processors without interruption, yielding a frayed pattern in the area $[1.8, 2.2] \times [0, 0.4]$ due to the distribution of the LTS clusters. By contrast, the ParMETIS/METIS framework yields a more homogeneous distribution in this case. We consider the DMR test case presented in Section 5.5.4 (adapted from Weber, 2018, Figure 4.5).

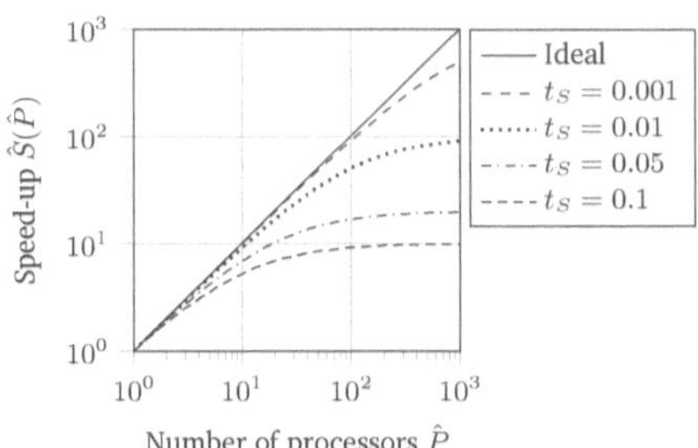

Figure 4.9: Theoretical speed-up for several sequential parts on multiple processors calculated by Amdahl's law (Amdahl, 1967). We assume that the run-time of a computational algorithm consists of a sequential and a parallel part ($T = t_S + t_P = 1$). The speed-up is calculated by $\hat{S}(\hat{P}) = T(1)/T(\hat{P}) = 1/(t_S + (1 - t_S)/\hat{P})$, where T denotes the entire run-time and $\hat{P}$ is the number of processors (adapted from Weber, 2018, Figure 4.7).

A speed-up study is performed on the Hessischer Hochleistungsrechner (HHLR) (*Lichtenberg*) using the DMR test case depicted in Figure 4.8. In general, if the number of degrees of freedom (DOF) per processor falls below a certain limit, the communication overhead prevents any further speed-up. This minimum number of DOF is problem-dependent (Hindenlang, 2014). On the HHLR, we use 5000 DOF/variable per processor. In total, we execute three runs for each number of processors $\hat{P} = \{1, 2, 4, 8, 16, 32\}$. For each configuration, we compute the arithmetic mean of the run-times[8]. We show the results of the speed-up study in Figure 4.10. ParMETIS/METIS focuses on the minimum of inter-processor communication, which becomes a bottleneck if the load per processor is lowered below a certain architecture-specific limit (Hindenlang, 2014). By contrast, the implemented Hilbert algorithm focuses on an even distribution of the computational load among all processors. As a result, the grid-partitioning based on Hilbert curves performs superior in this case.

In summary, Figure 4.10 shows that satisfactory results in terms of a good speed-up could not be reached. It seems that there is no general simple grid-partitioning technique for an efficient parallelization of LTS schemes. As an example, we imagine a processor border between two adjacent cells, which belong to different LTS cell clusters on different processors, $K_1 \in M_1$ on $\hat{P}_1$ and $K_2 \in M_2$ on $\hat{P}_2$ with the time-step sizes $\Delta t_2 \ll \Delta t_1$. Depending on the current grid-partitioning, which is strongly problem-dependent, there will be scenarios where $\hat{P}_1$ will have to wait for $\hat{P}_2$ until cell K_2 reaches the next synchronization level. This results in large idle times for processor $\hat{P}_1$. During our numerical experiments, we have observed this issue in multiple cases. However, we have shown that the grid-partitioning based on Hilbert curves is more suitable compared to the one based on the ParMETIS/METIS framework in the context of an adaptive LTS scheme.

---

[8]The run-time is measured using the *Trace Class* of the Microsoft .NET Framework (namespace *System.Diagnostics*). In the *Compressible Navier-Stokes (CNS)* solver of the *BoSSS* framework, the run-times of the methods *RunSolverOneStep* and *RedistributeAndMeshAdapt* are measured, including all sub-calls. These methods contain the evaluation of the spatial operator and the time-stepper as well as the grid-partitioning algorithm.

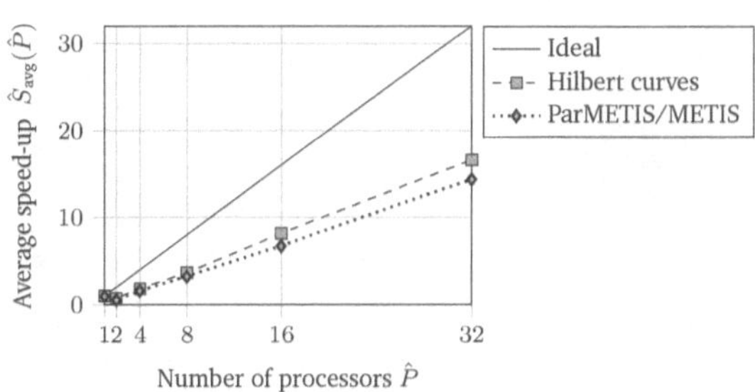

Figure 4.10: Average speed-up of the adaptive LTS scheme. The grid-partitioning based on Hilbert curves has a larger average speed-up than the one based on ParMETIS/METIS. The Hilbert algorithm focuses on an even distribution of the computational load, whereas ParMETIS/METIS tries to minimize the inter-processor communication in the first place and the load balance in the second. This only becomes important for a large number of processors (adapted from Weber, 2018, Figure 4.8).

5 Shock-Capturing on Cut-Cells

This chapter deals with a novel extension of a two-step shock-capturing strategy for the usage on an agglomerated cut-cell grid. The two-step shock-capturing strategy consists of a modal-decay detection step and an artificial viscosity smoothing step. Additionally, we apply an adaptive local time-stepping (LTS) scheme for computational speed-up. To the best of our knowledge, we were the first to publish such a combined approach in the context of a discontinuous Galerkin immersed boundary method (DG IBM) (Geisenhofer et al., 2019).

In Section 5.1, we outline the state of the art of classical shock-capturing strategies. We introduce several shock indicators and smoothing procedures in Sections 5.2 and 5.3, respectively. In Section 5.4, we discuss the applied two-step shock-capturing strategy and its extension to a DG IBM in detail. We present numerical results to verify this combined approach in Section 5.5, before closing the chapter with a brief summary in Section 5.6.

Figure 5.1 gives an illustrative overview of the numerical investigations and the underlying test cases of this chapter, in particular of Sections 5.4 and 5.5. First, we verify the basic properties of the two-step shock-capturing strategy, such as the $1/P^4$-scaling of the indicator value and the h/P-scaling of the maximum artificial viscosity parameter. Second, we consider several test cases, the Sod shock tube (Sod, 1978), a shock-vortex interaction, and a double Mach reflection (DMR), in order to verify the implementation in the *Compressible Navier-Stokes (CNS)* solver of the *BoSSS* framework in terms of solution accuracy. Additionally, we apply the adaptive LTS scheme presented in Chapter 4 and investigate the computational savings.

The idea behind shock-capturing　　Depending on the initial conditions, solutions of partial differential equations (PDEs) can be discontinuous. Examples of such PDEs range from the nonlinear scalar Burgers' equation to the nonlinear system of the Euler equations (2.5) for inviscid compressible flow. A common numerical treatment defining the class of shock-capturing strategies is to add artificial dissipation in order to obtain a stable solution of PDE systems. The addition of artificial viscosity changes the underlying governing equation and, thus, its numerical solution. We can think of this modified numerical solution as an inexact solution to the exact governing equation or as an exact solution to an inexact governing equation, following the derivations by Johnson et al. (1990) and Barter (2008). Here, the use of the expressions *exact numerical solution* implies that the numerical solution is converged and accurate up to machine accuracy. The latter point of view classifies the discrete numerical solution as the exact solution to a slightly perturbed PDE, called the *modified equation*. This modified equation can contain additional second, third or higher derivatives in the state variables, which affect the numerical solution differently, for example, by means of diffusion or dispersion effects.

The truncation error of first-order solutions contains second derivatives in the state variables. This means that the discontinuities in the solution of the modified equation are smoothed out

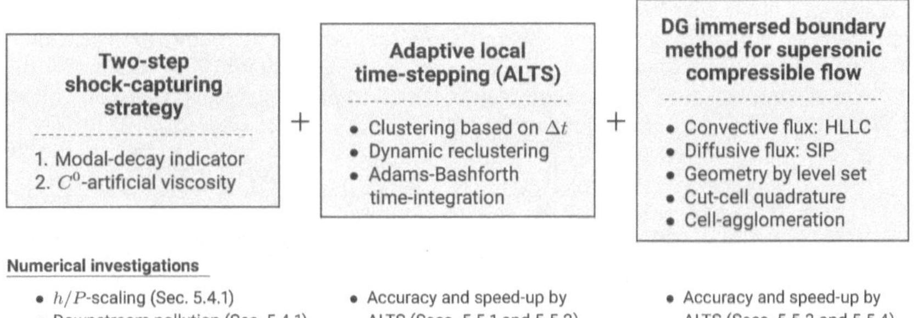

Figure 5.1: Overview of the numerical approaches applied in Chapter 5. The results are compared to benchmarks from the literature in Sections 5.4 and 5.5, focusing on different aspects of the interaction of the numerical approaches (gray boxes). The considered test cases are listed directly below.

by these terms. However, high-order solutions contain too little natural dissipation, since the magnitude of the inter-element jumps and therefore the added dissipation is proportional to $O(h^{P+1})$ (Persson and Peraire, 2006). Many high-order solutions contain third derivatives in their modified equations, leading to the effect of dispersion, which can be interpreted as a phase error in the higher modes. Dispersion will lead to oscillations centered around the discontinuity, since a discontinuity contains energy in all modes. These errors can spread throughout the entire computational domain, destroying the global high-order accuracy. Thus, it is a challenging issue to find a solution which is physically reasonable in either way, balancing a low accuracy and a nonphysical behavior. Some shock-capturing strategies solve this challenge by *locally* adding artificial dissipation, which numerically damps the oscillations in the vicinity of discontinuities.

5.1 State of the Art

When handling complex realistic flow situations such as high Mach number flows with discontinuous flow phenomena, discontinuous Galerkin (DG) methods are still susceptible to stability issues caused by under-resolution or oscillating polynomial solutions in the vicinity of discontinuities. To overcome this problem, several approaches were developed, which we briefly present in the following paragraphs based on the works by LeVeque (1999), Barter (2008), Toro (2009), and Geisenhofer et al. (2019).

Limiters Total variation diminishing (TVD) methods are commonly used when dealing with spurious oscillations in high-order methods, see Toro (2009, Section 13.1) and references therein. Originally developed for one-dimensional scalar problems, the basic idea of TVD methods can be traced back to the early works by Godunov (1959), Boris and Book (1973), and van Leer (1973). When considering the one-dimensional scalar transport equation $\partial\psi/\partial t = u_1\,\partial\psi/\partial x_1$, the continuous total variation (TV) is defined as

$$\text{TV}(\psi(x_1, t)) = \int \left|\frac{\partial\psi}{\partial x_1}\right| \mathrm{d}x_1\,. \tag{5.1}$$

A numerical method is said to be total variation diminishing (TVD) if the discrete version of Equation (5.1) satisfies the condition

$$\text{TV}(\psi(x_1, t_{n+1})) \leq \text{TV}(\psi(x_1, t_n))\,. \tag{5.2}$$

TVD methods can be divided into *flux limiters* and *slope limiters*. In the case of flux limiters, the limiting procedure acts on the numerical flux, whereas in the case of slope limiters, the limiter acts on the solution itself. The latter kind of methods reduces the (oscillating) gradient in a cell by additionally considering the solution in the neighboring cells (van Leer, 1973). An extension of the TVD approach to DG methods was developed by Cockburn and Shu (1989), Cockburn et al. (1989), Cockburn et al. (1990), and Cockburn and Shu (1991) using the *minmod*-operator. Thereby, the polynomial solution is reduced to a zeroth-order representation in regions where the limiter is active. An inherent shortcoming of this approach is that the convergence to the steady-state solution is sometimes hard to obtain (Barter, 2008).

Another sub-class of limiters are *a-posteriori limiters* (Giani et al., 2014). For example, Dumbser et al. (2014) recomputed the numerical solution in troubled cells using a sub-cell grid procedure based on the Multi-Dimensional Optimal Order Detection (MOOD) approach, which originates from the Finite Volume Method (FVM). The main idea of the MOOD approach is to obtain a numerical candidate solution, which is tested against some admissibility criteria, by calculating an 'unlimited' time-step. Troubled cells are identified and the solution in these cells is reconstructed by a low-order solution following the original MOOD approach. By contrast, the version of Dumbser et al. (2014) maintains the high-order accuracy without losing the sub-cell resolution property of the DG method.

Reconstruction methods In contrast to limiting procedures, reconstruction methods use the high-order modes of the polynomial solution for resolving discontinuities in a sharp manner. Harten and Osher (1987) introduced the essentially non-oscillatory (ENO) scheme, which was later extended by Shu and Osher (1988). In general, ENO schemes choose a stencil of the piece-wise constant solutions in the neighboring cells in order to reconstruct a non-oscillating high-order solution. Liu et al. (1994) extended the idea by means of the weighted essentially non-oscillatory (WENO) scheme, where multiple stencils are constructed and then combined in a nonlinear weighted way.

For an introduction to ENO/WENO schemes in the context of the DG method, we refer to the overview by Shu (2016). A shortcoming of the classical WENO approach is that the construction of large stencils, which incorporate the solution of neighboring cells, destroys the locality of the DG method. As a remedy, Qiu and Shu (2004, 2005) developed a scheme using Hermite polynomials in order to keep a compact stencil on structured grids. Compared to the original

WENO scheme, their adaption uses the information from the derivatives of the solution to maintain the locality of the stencil. Consequently, the size of the stencil is smaller when aiming for the accuracy as that of the original scheme. Furthermore, Luo et al. (2007) published an extension to unstructured grids. Similar to the disadvantage of limiters in the context of the DG method, it is hard to find steady-state solutions when using WENO schemes, rendering them not well suited for the application of implicit time-stepping schemes (Barter, 2008).

Artificial viscosity methods[9]   While the previously introduced approaches have their advantages, the simplicity of the artificial viscosity approach, originally introduced by Von Neumann and Richtmyer (1950), remains appealing to this day. Further larger adaptions were made by Baldwin and MacCormack (1975) and Jameson et al. (1981). The basic idea is to add a second-order diffusive term to the governing equations in order to smooth discontinuities over a layer where they can be adequately resolved by the numerical scheme and, thus, prevent undesired oscillations.

Bey and Oden (1991, 1996) along with Bassi and Rebay (1997, 2000) already demonstrated the potential of the artificial viscosity approach for high-order DG methods. However, determining the 'correct' amount of artificial viscosity across a large range of parameters and flow configurations has remained a delicate issue. Persson and Peraire (2006) proposed a simple solution for this issue by using the modal-decay of the DG coefficients (Krivodonova et al., 2004) as a robust indicator for the smoothness of the solution. In the approach of Persson and Peraire (2006), the artificial viscosity and the shock thickness scale with the DG resolution $O(h/P)$ so that the shock can be resolved within one computational cell for a large enough resolution, which is commonly called *sub-cell accuracy*.

Overcoming the problems of a cell-wise constant artificial viscosity, which introduces additional oscillations in the vicinity of the discontinuities, Barter and Darmofal (2007, 2010) published a new high-order state-based artificial viscosity formulation by means of a PDE. Persson (2013) extended his earlier work (Persson and Peraire, 2006) with the introduction of an at least C^0-continuous artificial viscosity field formulation in order to deal with the aforementioned stability issues. Other authors analyzed the performance of the modal-decay indicator and improved the basic version by introducing additional constraints (Klöckner et al., 2011) or developing a dynamic threshold setting (Lv et al., 2016).

5.2 Shock Detection

Shock indicators are an essential ingredient for shock-capturing strategies based on artificial viscosity, since they quantify the smoothness of numerical solutions and, thus, can be used to determine the required amount of artificial viscosity.

In the literature, there are many popular approaches such as the ones considering the L^2-norm of the residual of the solution in variational form (Bassi and Rebay, 1995; Jaffre et al., 1995). Based on this, Hartmann (2006) developed a similar approach by detecting the orientation of the discontinuity and refining the grid anisotropically. Another approach uses local information over cell edges or faces (Bassi et al., 1997). Further developments consider the jumps of the

---

[9]Modified version of Geisenhofer et al. (2019, Section 1).

solution across cell boundaries in order the evaluate the smoothness and the convergence of the solution (Krivodonova et al., 2004; Feistauer and Kučera, 2007; Barter and Darmofal, 2010). A different variant employs an entropy-based nonlinear viscosity model for Fourier approximations of conservation laws (Guermond and Pasquetti, 2008).

Due to its simplicity, another popular approach is the modal-decay indicator, which uses the local L^2-norm of the discretized solution (Persson and Peraire, 2006). The previously mentioned works all lack a suitable scaling under h/P-refinement. In analogy to the decay rate of a Fourier expansion, Persson and Peraire (2006) solved this scaling issue and showed that their indicator variant decays with a rate of $1/P^4$. In order to understand the working principle, we can think of the superposition of multiple sine waves. The magnitudes of the higher modes correspond to the high frequency parts and are therefore responsible for the oscillating behavior of the numerical solution. A further development was published by Klöckner et al. (2011), who presented, among others, a detailed analysis of the original modal-decay variant and showed its limiting cases. For a detailed overview and discussion of different shock indicators, we refer the reader to the works by Klöckner et al. (2011) and Lv et al. (2016).

5.2.1 Modal-Decay Indicator

In this work, we use a variant of the modal-decay shock indicator as proposed by Persson and Peraire (2006). We introduce the generic variant for the application on a non-agglomerated cut-cell grid $\mathcal{K}_h^X$ of a DG IBM, since it incorporates the standard version on the uncut background grid $\mathcal{K}_h$ and the agglomerated grid $\mathcal{K}_h^{X,\mathrm{agg}}$, see Definitions 3.1, 3.4 and 3.8, respectively.

Strategy[10]   First, we write the cell-local polynomial solution (3.7) in a cell $K_j \in \mathcal{K}_h^X$, $K_j \neq \emptyset$, as

$$\psi_j(\mathbf{x}, t) := \psi|_{K_j} \in \mathbb{P}_P^X(\{K_j\}). \tag{5.3}$$

Second, we introduce a truncated solution using a basis of order $P - 1$

$$\hat{\psi}_j(\mathbf{x}, t) = \Pi^{P-1}(\psi_j), \tag{5.4}$$

by using the projection operator

$$\Pi^{P-1} : \mathbb{P}_P^X(\{K_j\}) \to \mathbb{P}_{P-1}^X(\{K_j\}),$$
$$\psi \mapsto \hat{\psi}, \tag{5.5}$$

with the essential property $\langle \psi - \hat{\psi}, \vartheta \rangle = 0$, $\forall \vartheta \in \mathbb{P}_{P-1}^X(\{K_j\})$. Following Persson and Peraire (2006), we define a cell-local modal-decay shock indicator as

$$S_j(t) = \frac{\langle \psi_j - \hat{\psi}_j, \psi_j - \hat{\psi}_j \rangle}{\langle \psi_j, \psi_j \rangle}. \tag{5.6}$$

For smooth solutions $\psi_j(\mathbf{x}, t) \in C^0$, we can expect the coefficients of the polynomial expansion to decay quickly. For non-smooth solutions, the magnitude of the discontinuity or the

---

[10]Modified version of Geisenhofer et al. (2019, Section 4.1).

extent of the under-resolution prescribe the rate of decay. In analogy to spectral analysis, the coefficients of a one-dimensional Fourier expansion decay at least with a rate of $1/P^2$ for smooth solutions (Mavriplis, 1994). Assuming the same behavior for the polynomial expansion and considering that Equation (5.6) contains squared quantities, Persson and Peraire (2006) confirmed through numerical experiments that the indicator value decays with a rate of $1/P^4$. In this work, we use the discrete density field as the input variable for determining the indicator values, since we solve for the conservative variables (2.6). When solving for the primitive variables, the pressure is often used as an input for the indicator as well.

In particular, if we consider an orthogonal cell-local basis $\phi_{j,n}(\mathbf{x})$, we can write the full solution (5.3) and the truncated solution (5.4) as

$$
\psi_j(\mathbf{x}, t) = \sum_{n=1}^{N(P)} \tilde{\psi}_{j,n}(t)\, \phi_{j,n}(\mathbf{x}), \qquad \mathbf{x} \in K_j, \tag{5.7a}
$$

$$
\hat{\psi}_j(\mathbf{x}, t) = \sum_{n=1}^{N(P-1)} \tilde{\psi}_{j,n}(t)\, \phi_{j,n}(\mathbf{x}), \qquad \mathbf{x} \in K_j, \tag{5.7b}
$$

respectively, where $\tilde{\psi}_{j,n}(t)$ denote the coefficients and $N(P)$ is the number of terms in the polynomial expansion of order P. We insert Equation (5.7) into Equation (5.6), yielding

$$
S_j(t) = \frac{\sum_{n=N(P-1)}^{N(P)} \sum_{m=N(P-1)}^{N(P)} \tilde{\psi}_{j,n} \tilde{\psi}_{j,m} \int \phi_{j,n}\phi_{j,m}\, \mathrm{d}V}{\sum_{n=1}^{N(P)} \sum_{m=1}^{N(P)} \tilde{\psi}_{j,n} \tilde{\psi}_{j,m} \int \phi_{j,n}\phi_{j,m}\, \mathrm{d}V}, \tag{5.8}
$$

where the integral $\int \phi_{j,n}\phi_{j,m}\, \mathrm{d}V$ corresponds to the cell-local mass matrix $\mathbf{M}_j$, see also Equation (3.18). The structure of $\mathbf{M}_j$ strongly depends on the shape of the cut-cells and is different from the mass matrix in uncut cells. In these cells, the mass matrix equals the identity matrix ($\mathbf{M}_j = \mathbf{I}$), since the basis $\phi_{j,n}(\mathbf{x})$ is orthogonal. Consequently, Equation (5.8) reduces to

$$
S_j(t) = \frac{\sum_{n=N(P-1)}^{N(P)} \tilde{\psi}_{j,n}^2}{\sum_{n=1}^{N(P)} \tilde{\psi}_{j,n}^2}. \tag{5.9}
$$

In the remainder of this work, we use the presented modal-decay indicator in its formulations (5.8) and (5.9) as the input variable for the calculation of artificial viscosity due to its favorable properties such as the h/P-scaling and the straightforward extension for the use on an agglomerated cut-cell grid $\mathcal{K}_h^{\mathrm{X,agg}}$. We would like to anticipate that there was no limitation in the applicability or performance of this indicator on $\mathcal{K}_h^{\mathrm{X,agg}}$. Furthermore, we do not focus on the improvements or other variants (Barter and Darmofal, 2010; Klöckner et al., 2011; Lv et al., 2016), but we rather apply the previously introduced version to demonstrate the applicability in the context of a DG IBM.

Analysis A detailed analysis of the modal-decay indicator (Persson and Peraire, 2006) can be found in the work by Klöckner et al. (2011). We state the most relevant findings here.

For convenience, we repeat the notation of the cell-local approximation $\psi_j(\mathbf{x}, t)$ in a cell K_j as

$$
\psi_j(\mathbf{x}, t) = \sum_{n=1}^{N} \tilde{\psi}_{j,n}(t)\, \phi_{j,n}(\mathbf{x}) = \tilde{\psi}_j(t) \cdot \phi_j(\mathbf{x}), \qquad \mathbf{x} \in K_j, \tag{3.7, repeated}
$$

where $\tilde{\psi}_j = (\tilde{\psi}_{j,n})_{n=1,\ldots,N}$ denotes the time-dependent coefficient vector of a modal basis $\phi_j = (\phi_{j,n})_{n=1,\ldots,N} \in \mathbb{P}_P(\mathcal{K}_h)$, compare Equations (3.1), (3.6) and (3.8). Klöckner et al. (2011) assume that monotone modal-decay can be approximately represented by the relation

$$\left|\tilde{\psi}_{j,n}\right| \sim Cn^{-s_\mathrm{d}}, \tag{5.10}$$

where s_d is the decay rate and C is a constant. Taking the logarithm of Equation (5.10) results in

$$\log\left|\tilde{\psi}_{j,n}\right| \sim \log(C) - s_\mathrm{d}\log(n). \tag{5.11}$$

The terms $\log(C)$ and s_d of the affine relation (5.11) can be determined by a least-squares fitting with the condition

$$\sum_{n=2}^{N}\left|\log\left|\tilde{\psi}_{j,n}\right| - [\log(C) - s_\mathrm{d}\log(n)]\right|^2 \rightarrow \min. \tag{5.12}$$

If the cell-local solution $\psi_j(\mathbf{x},t)$ (3.7) is known, Equation (5.12) yields an expression for the decay rate s_d. In analogy to spectral analysis, Klöckner et al. (2011) expect the following decay rates for different discontinuous solutions:

$$\begin{aligned}
s_\mathrm{d} &\approx 1, && \text{for discontinuous } \psi_j(\mathbf{x},t), \\
s_\mathrm{d} &\approx 2, && \text{for } \psi_j(\mathbf{x},t) \in C^0 \setminus C^1, \\
s_\mathrm{d} &\approx 3, && \text{for } \psi_j(\mathbf{x},t) \in C^1 \setminus C^2.
\end{aligned}$$

They investigate several test cases in order to verify the expected decay rates:

- *Heaviside function* $\psi_j(x_1,t) = H(x_1)$: The expectation is met well, delivering a decay rate of $s_\mathrm{d} = 1$.

- *Kink function* $\psi_j(x_1,t) = x_1 H(x_1)$: A decay rate of $s_\mathrm{d} = 7.2$ is obtained. This setting clearly shows a shortcoming of the modal-decay indicator. The magnitudes of the odd coefficients $|\tilde{\psi}_{j,n}|$ of number three and greater are zero due to an odd-even pattern. Coefficients with zero magnitude indicate too much smoothness in the respective modes, leading to an overestimated smoothness. This is due to the fact that assumption (5.10) only models monotone decay, whereas the kink function is characterized by a strongly non-monotone mode pattern. In order to overcome the problem caused by the odd-even pattern, Klöckner et al. (2011) propose a modification, called *skyline pessimization*, which detects spurious modes. From the magnitudes of the original modal coefficients $|\tilde{\psi}_{j,n}|_{n=1,\ldots,N}$, a new set of modal coefficients is created by

$$\tilde{\psi}_{j,n}^{\mathrm{sl}} := \max_{k \in \{\min(n,N-1),\ldots,N)\}} \left|\tilde{\psi}_{j,n}\right|, \qquad n = 2,\ldots,N, \tag{5.13}$$

where the small spurious coefficients are removed from the fit. The skyline pessimization approach delivers a corrected decay rate of $s_\mathrm{d} \approx 1.67$, close to the exact rate of $s_\mathrm{d} = 2$.

- *Truncated polynomial* $\psi_j(x_1,t) = x_1^2 H(x_1)$: For $\psi_j(\mathbf{x},t) \in C^1 \setminus C^2$, the magnitudes of the coefficients contain an odd-even pattern similar to the kink function case. The indicator delivers an overestimated decay rate of $s_\mathrm{d} \approx 13$, corrected by skyline pessimization to $s_\mathrm{d} \approx 3$.

- *Constant function perturbed by white noise:* Since the indicator optimized by skyline pessimization, see Equation (5.13), does not consider the constant mode information, it delivers a decay rate of $s_{\mathrm{d}} = 0$. In order to eliminate this undesired effect, the information of the constant value is re-added by distributing the energy according to a 'perfect modal-decay', which Klöckner et al. (2011) denote as *baseline modal-decay* with the coefficients $\tilde{\psi}_{j,n}^{\mathrm{bd}}$, given by

$$\left| \tilde{\psi}_{j,n}^{\mathrm{bd}} \right| \sim \frac{1}{\sqrt{\sum_{n=2}^{N} \frac{1}{n^{2P}}}} \frac{1}{n^{P}} \,, \tag{5.14}$$

while ensuring

$$\sum_{n=2}^{N} \left| \tilde{\psi}_{j,n}^{\mathrm{bd}} \right|^{2} = 1 \,. \tag{5.15}$$

Additionally, they consider the new coefficients

$$\underbrace{\left| \tilde{\psi}_{j,n}^{*} \right|^{2}}_{\substack{\text{input for skyline} \\ \text{pessimization}}} := \underbrace{\left| \tilde{\psi}_{j,n} \right|^{2}}_{\substack{\text{original} \\ \text{coefficients}}} + \underbrace{\| \psi_{j}(\mathbf{x}, t) \|_{L^{2}(K_{j})}^{2}}_{\substack{\text{local} \\ L^{2}-\text{norm}}} \underbrace{\left| \tilde{\psi}_{j,n}^{\mathrm{bd}} \right|^{2}}_{\substack{\text{baseline modal} \\ \text{decay}}} \,, \qquad n = 2, \dots, N \,, \tag{5.16}$$

as a more suitable input for the skyline pessimization procedure after adding the baseline modal-decay scaled by the L^{2}-norm of the local solution $\psi_{j}(\mathbf{x}, t)$.

The work by Klöckner et al. (2011) clearly improves the basic version of the modal-decay indicator with reasonably larger computational costs. However, we apply the original version by Persson and Peraire (2006) in this work due to the following reasons:

- The basic modal-decay indicator is mathematically simple and easy to implement.

- The modified version of Klöckner et al. (2011) does not solve situations where a strong discontinuity is located close to a cell boundary. In these cases, the smoothness estimates can differ up to two orders of magnitude from the correct decay rate. During our numerical experiments, we observed a similar behavior of the basic version, since the zeroth-order coefficient has a strong influence on the final indicator value, for example, depending on whether the discontinuity is located close to the left or the right cell boundary. This scenario strongly influences the local mean value of the numerical solution and, thus, the indicator value.

- We consider the C^{0}-projection of the artificial viscosity field, which we present in Section 5.3.1, to be a remedy for the drawbacks of the modal-decay indicator, since it levels the problems mentioned above.

5.2.2 Jump Indicator

Another type of indicator is based on the evaluation of the inter-element jumps of the numerical solution. The original version of the jump indicator was published by Dolejší et al. (2003) and later adapted by Krivodonova et al. (2004). In the following, we briefly introduce the jump indicator based on the notation by Barter (2008). For a DG method, the magnitude of the inter-element jumps should converge as

$$
|[\![\psi(\mathbf{U})]\!]| = \begin{cases} O(h^{P+1}), & \text{for smooth solutions}, \\ O(1), & \text{for discontinuous solutions}, \end{cases} \tag{5.17}
$$

where $\psi = \psi(\mathbf{U})$ is a component of the state vector $\mathbf{U}$, see Equation (2.6), or a derived quantity. Based on this relation, Barter (2008) defines the cell-local jump indicator as

$$
\tilde{J}_j(\psi_j, t) = \frac{1}{|\partial K_j|} \oint_{\partial K_j} \left| \frac{[\![\psi_j]\!]}{\{\!\{\psi_j\}\!\}} \right| \, \mathrm{d}S, \tag{5.18}
$$

which relates the jump of some quantity $[\![\psi_j]\!]$ to the average jump $\{\!\{\psi_j\}\!\}$ across a cell boundary ∂K_j, see also Equations (3.2) to (3.4). The modal-decay indicator is based on the solution in a single cell, whereas the jump indicator additionally requires the evaluation of the solution in neighboring cells. Depending on the chosen artificial viscosity approach, the usage of the jump indicator can destroy the compactness of the numerical scheme. This happens if the solutions of second-degree neighboring cells are needed for the evaluation of the integral in Equation (5.18). In this work, we abstain from the usage of the jump indicator due to its performance problems in the context of the Euler equations (2.5) (Lv et al., 2016).

5.3 Shock Smoothing With Artificial Viscosity

As described in Section 5.2, a shock indicator provides a numerical value depending on which a local smoothing procedure may be activated. In general, an artificial viscosity approach smooths the numerical solution in the vicinity of discontinuities by adding an additional second-order diffusive term to the original equations. In the case of the Euler equations (2.5), we obtain

$$
\underbrace{\frac{\partial \mathbf{U}}{\partial t} + \frac{\partial \mathbf{F}_1(\mathbf{U})}{\partial x_1} + \frac{\partial \mathbf{F}_2(\mathbf{U})}{\partial x_2}}_{\text{Euler equations}} = \underbrace{\varepsilon \left(\frac{\partial^2 \mathbf{U}}{\partial x_1^2} + \frac{\partial^2 \mathbf{U}}{\partial x_2^2} \right)}_{\substack{\text{artificial viscosity} \\ \text{term}}}, \tag{3.50, repeated}
$$

where ε is the artificial viscosity parameter, which we will determine in Section 5.3.1. Note that we use isotropic artificial viscosity, which is a special case of the general anisotropic formulation. Artificial viscosity spreads the discontinuities over a length scale which can be adequately resolved by the numerical scheme. In general, we want to add as little artificial viscosity as possible and only locally at the discontinuities. We also enforce a scaling dependency on the resolution $O(h/P)$ according to Persson and Peraire (2006).

5.3.1 Laplacian Artificial Viscosity[11]

In this work, we apply a Laplacian artificial viscosity approach and enforce the artificial viscosity field to be at least C^0-continuous (Persson, 2013). In the case of oblique discontinuities, a piece-wise constant artificial viscosity formulation misses the point of the smoothing idea, since

---

[11]Modified version of Geisenhofer et al. (2019, Section 4.2).

additional oscillations are introduced. From a numerical point of view, this can be explained by the drastic change of the type of the equation between neighboring cells having zero/low and high artificial viscosity values, respectively. Consequently, Equation (3.50) is not strictly hyperbolic anymore in cells where artificial viscosity is added.

Preliminary cell-local artificial viscosity   For determining the preliminary cell-local amount of artificial viscosity, we use a smoothed Heaviside function

$$\forall \mathbf{x} \in K_j : \qquad \varepsilon'_j(\mathbf{x}, t) = \begin{cases} 0, & \text{if } s_j < s_0 - \kappa \\ \frac{\varepsilon_0}{2} \left(1 + \sin \frac{\pi(s_j - s_0)}{2\kappa} \right), & \text{if } s_0 - \kappa \leq s_j \leq s_0 + \kappa, \\ \varepsilon_0, & \text{if } s_j > s_0 + \kappa, \end{cases} \qquad (5.19)$$

where $s_j = \log(S_j)$, see also Equations (5.8) and (5.9). Persson and Peraire (2006) add scaling factors to the maximum viscosity value $\varepsilon_0 \sim h/P$ and the pre-defined indicator value $s_0 \sim \log(1/P^4)$ in Equation (5.19). As soon as a troubled cell is detected, artificial viscosity is added depending on the strength of the shock. Some authors (Klöckner et al., 2011) speculated about a typographical error in the original work (Persson and Peraire, 2006), since the logarithm of the indicator value should not scale with $s \sim 1/P^4$ but rather with $s \sim \log(1/P^4)$. In the numerical experiments of this work, the original scaling $s \sim 1/P^4$ can be confirmed. The parameter $\kappa \sim O(1)$ determines the width of the smoothed Heaviside profile and has to be adequately adjusted in order to obtain a sharp shock profile. In this work, we use values of $\kappa = \{0.5, 1.0\}$ which are determined by numerical testing.

Scaled cell-local artificial viscosity   The added amount of artificial viscosity additionally depends on some local characteristic velocity λ_c. Klöckner et al. (2011) started from investigating the fundamental solution of the diffusion equation $\partial u_1/\partial t = \varepsilon \Delta u_1$ and derived the scaling expressions $\varepsilon_0 \sim \lambda_c$ and $\varepsilon_0 \sim h/P$, verifying the results of Barter and Darmofal (2010). Several characteristic velocities, such as the supersonic propagation speed of a shock wave or the sonic background propagation velocity of a contact discontinuity, could be a natural choice for λ_c. The question remains which velocity to choose. Klöckner et al. (2011) stated that it is reasonable to take the maximum local characteristic velocity $\lambda_{c,\max} = |\mathbf{u}| + a$, corresponding to that of a shock wave, in order to get a running implementation with the price of smoothing weak features, such as contact discontinuities, more than necessary. Furthermore, they empirically derived a correction factor of $1/2$ for the Euler equations (2.5). Based on this, we determine the piece-wise cell-local amount of artificial viscosity from the expression

$$\varepsilon_j(\mathbf{x}, t) = \frac{1}{2} \, \varepsilon'_j(\mathbf{x}, t) \, \lambda_{c,\max} \frac{h}{P} \, . \qquad (5.20)$$

Boundary condition   During the course of this work, we tested several boundary conditions in the artificial viscosity flux implementation, which is based on a standard symmetric interior penalty (SIP) formulation (Arnold, 1982; Arnold et al., 2002), see Section 3.2.3 for details. As this work focuses on the applicability in the context of a DG IBM, we abstain from investigating this issue in excessive detail. We achieve best results in terms of stability and accuracy when using zero-Neumann boundary conditions, which we apply in the remainder of this work.

C^0**-continuous artificial viscosity**   There are several publications (Barter and Darmofal, 2010; Klöckner et al., 2011; Lv et al., 2016) about the advantages and drawbacks of the original approach (Persson and Peraire, 2006), each of them having their own improvements. We stick to the standard approach except for the modifications presented above. In any event, we would like to mention that it is essential to enforce C^0-continuity of the artificial viscosity field (5.20) for a robust and reliable method (Barter and Darmofal, 2010; Persson, 2013). In order to obtain a final C^0-continuous formulation of the artificial viscosity field, we define a piece-wise linear continuous space

$$\mathbb{Q}_1(\mathcal{K}_h) := \left\{ f \in L^2(\Omega) : \forall K_j \in \mathcal{K}_h : f|_{K_j \cap \Omega} = \sum_{0 \le k,l \le 1} x^k y^l b_{kl}, \ b_{kl} \in \mathbb{R} \right\} \cap C^0(\Omega), \qquad (5.21)$$

with a maximum polynomial degree of $P = 1$. We define the corresponding projection operator

$$\begin{aligned} \Pi^{C^0} : L^2(\Omega) &\to \mathbb{Q}_1(\mathcal{K}_h), \\ \psi &\mapsto \psi^{C^0}, \end{aligned} \qquad (5.22)$$

with the essential property $\langle \psi - \psi^{C^0} | \vartheta \rangle = 0$, $\forall \vartheta \in \mathbb{Q}_1$. Finally, we project the global piece-wise constant artificial viscosity field, see the local form in Equation (5.20), onto the space $\mathbb{Q}_1$ (5.21), via

$$\varepsilon^{C^0}(\mathbf{x}, t) = \Pi^{C^0}(\varepsilon(\mathbf{x}, t)), \qquad (5.23)$$

in order to enforce continuity. The shock indicator (5.6) and the artificial viscosity field (5.23) are updated in each (local) time-step. When using the LTS scheme presented in Chapter 4, the update is performed at every synchronization level.

Note on the implementation   In general, the preliminary scaled piece-wise constant artificial viscosity field $\varepsilon(\mathbf{x}, t)$ has jumps across the cell boundaries by construction. This is the case when using the background grid $\mathcal{K}_h$, the cut-cell grid $\mathcal{K}_h^{\mathsf{X}}$ or the agglomerated grid $\mathcal{K}_h^{\mathsf{X,agg}}$. Subsequently, we project the piece-wise constant discontinuous artificial viscosity field $\varepsilon(\mathbf{x}, t)$ onto the piece-wise linear continuous space $\mathbb{Q}_1(\mathcal{K}_h)$ by using the projection $\Pi^{C^0}(\varepsilon(\mathbf{x}, t))$. The artificial viscosity field is realized on the background grid $\mathcal{K}_h$ with the corresponding polynomial space $\mathbb{P}(\mathcal{K}_h)$, whereas the spatial operator is evaluated on the non-agglomerated cut-cell space $\mathbb{P}(\mathcal{K}_h^{\mathsf{X}})$ and then projected onto the agglomerated space $\mathbb{P}(\mathcal{K}_h^{\mathsf{X,agg}})$, see Equations (3.69) and (3.71), respectively.

5.3.2 Physical Artificial Viscosity

In contrast to the Laplacian artificial viscosity approach presented in Section 5.3.1, the physical approach is based on the dissipation behavior of an ideal gas (Persson and Peraire, 2006). In this context, the Euler equations (2.5) are extended by generic viscous flux vectors $\mathbf{F}_1^v(\mathbf{U}) : \mathbb{R}^4 \to \mathbb{R}^4$ and $\mathbf{F}_2^v(\mathbf{U}) : \mathbb{R}^4 \to \mathbb{R}^4$ to

$$\frac{\partial \mathbf{U}}{\partial t} + \frac{\partial \mathbf{F}_1(\mathbf{U})}{\partial x_1} + \frac{\partial \mathbf{F}_2(\mathbf{U})}{\partial x_2} = \frac{\partial \mathbf{F}_1^v(\mathbf{U})}{\partial x_1} + \frac{\partial \mathbf{F}_2^v(\mathbf{U})}{\partial x_2}, \qquad (5.24)$$

with

$$\mathbf{F}_1^v(\mathbf{U}) = \begin{pmatrix} 0 \\ \tilde{\tau}_{11} \\ \tilde{\tau}_{21} \\ u_1\tilde{\tau}_{11} + u_2\tilde{\tau}_{21} - \hat{Q}_1 \end{pmatrix}, \qquad \mathbf{F}_2^v(\mathbf{U}) = \begin{pmatrix} 0 \\ \tilde{\tau}_{12} \\ \tilde{\tau}_{22} \\ u_1\tilde{\tau}_{12} + u_2\tilde{\tau}_{22} - \hat{Q}_2 \end{pmatrix}. \tag{5.25}$$

The stresses are modeled by

$$\tilde{\tau}_{ij} = \mu \left[\left(\frac{\partial u_i}{\partial u_j} + \frac{\partial u_j}{\partial u_i} \right) - \frac{2}{3} \delta_{ij} \frac{\partial u_k}{\partial x_k} \right], \qquad \text{for } i, j = 1, 2, \tag{5.26}$$

and the heat flux is defined by

$$\hat{Q}_i = -k \frac{\partial T}{\partial x_i}, \qquad \text{for } i = 1, 2. \tag{5.27}$$

Persson and Peraire (2006) determine the viscosity coefficient as a function of the temperature with $\mu = \mu^*\sqrt{T/T^*}$, where μ^* is the viscosity coefficient at the sonic temperature $T^* = 2T_0/(\gamma + 1)$ with temperature T_0 at stagnation pressure conditions.

Viscosity calculation Liepmann and Roshko (2001) showed that the shock thickness δ_s is proportional to the viscosity coefficient μ ($\delta_s \sim \mu$). By considering the change of the normal velocity across a shock wave $\Delta u = 2u_\infty(M_\infty^2 - 1)/([\gamma+1]M_\infty^2)$, compare Equation (2.30), as well as the values of the density $\rho^* = \rho\,(M = 1, T = T^*)$ and the viscosity $\mu^* = \mu\,(M = 1, T = T^*)$ at sonic conditions, Persson and Peraire (2006) state the expression

$$\frac{\delta_s \Delta u \rho^*}{\mu^*} \approx 1. \tag{5.28}$$

The sonic density ρ^* can be sufficiently approximated by the free-stream density ρ_∞ and Mach number M_∞. Persson and Peraire (2006) argue that the sonic viscosity μ^* has to be determined in way so that Equation (5.28) is fulfilled in order to resolve a shock with a thickness of $O(h/P)$. A relation between the viscosity and the thermal conductivity coefficient is given by the Prandtl number $Pr = \mu_\infty c_p/k_\infty$. Persson and Peraire (2006) distinguish between two different physical models for $Pr = 3/4$ and $Pr = 0$. In the first case, which is close to the Prandtl number in air, they used the entropy as the input variable for the shock indicator. In the second case, which neglects heat transfer effects, the shock indicator acts on the enthalpy field. After the shock indicator field has been evaluated, see Equations (5.8) and (5.9), the final viscosity is added according to the piece-wise constant version (5.19), where the sonic viscosity μ^* determines the maximum artificial viscosity ε_0 ($\varepsilon_0 = \mu^*$). We abstain from using this model due to its complexity, since no general significant superiority of the physical model over the Laplacian model is reported.

5.4 Two-Step Shock-Capturing Strategy

In this work, we use a two-step shock-capturing strategy consisting of a detection step and a smoothing step, see Sections 5.2 and 5.3. In the first step, we detect oscillating solutions by means of a modal-decay indicator (5.6) and its extended variant (5.8) for a DG IBM. In the second step, we apply a C^0-continuous artificial viscosity formulation (5.23).

5.4.1 Basic Properties

In this section, we discuss the basic properties of the two-step shock-capturing strategy by investigating the linear scalar transport equation and the nonlinear inviscid Burgers' equation.

Scalar transport equation We consider the one-dimensional linear scalar transport equation for some scalar quantity $\psi(x_1, t) : \mathbb{R} \to \mathbb{R}$, written in the form

$$\frac{\partial \psi(x_1, t)}{\partial t} + u_1 \frac{\partial \psi(x_1, t)}{\partial x_1} = 0 \,, \tag{5.29}$$

where the advection velocity is assumed to be $u_1 = 1$. We choose the discontinuous initial conditions

$$\psi_0(x_1) = \psi(x_1, 0) = \begin{cases} 0.5 \,, & \text{if } x_1 \leq 0.45 \,, \\ 1 \,, & \text{if } x_1 > 0.45 \,, \end{cases} \tag{5.30}$$

on a pseudo-two-dimensional computation domain $\Omega = [0, 1] \times [0, 1]$, being periodic in x_2-direction, with a characteristic grid size of $h = 0.1$ for a polynomial degree of $P = 2$. In this setting, the initial discontinuity at $x_1 = 0.45$ does not coincide with a cell boundary. We keep the complexity as low as possible for showing the basic properties so that we activate artificial viscosity by a simple on/off-switch

$$\varepsilon'_j(x_1, t) = \begin{cases} 0 \,, & \text{if } S_j \leq S_0 \,, \\ \varepsilon_0 \,, & \text{if } S_j > S_0 \,, \end{cases} \tag{5.31}$$

using a piece-wise constant version in contrast to Equation (5.23). We apply the explicit Euler scheme (4.7) for time-integration, where the maximum admissible time size Δt is calculated adaptively according to Equation (4.26).

This test case, being simple at first glance, is challenging in the context of shock-capturing strategies based on artificial viscosity, since the scalar transport equation (5.29) is not self-steepening in contrast to the Burgers' equation (5.34) or the Euler equations (2.5). As a result, an initial discontinuity will never return to its former shape, once it is smoothed by artificial viscosity even if shock-capturing is deactivated. Consequently, a comparably high h/P-resolution is needed in order to obtain sub-cell resolution.

The first row of Figure 5.2 shows the temporal evolution of the initial jump profile (5.30). The solid and the dashed lines show the profiles for a simulation with deactivated and activated shock-capturing, respectively. The solution oscillates if shock-capturing is deactivated, whereas the profile is smooth and spread over several cells if shock-capturing is activated. The second row of Figure 5.2 shows the indicator values if shock-capturing is activated. We set the indicator value for activating artificial viscosity to

$$S_0 = 6.25 \cdot 10^{-4} \frac{1}{P^4} \,, \tag{5.32}$$

and the maximum amount of artificial viscosity to

$$\varepsilon_0 = 10.0 \frac{h}{P} \,. \tag{5.33}$$

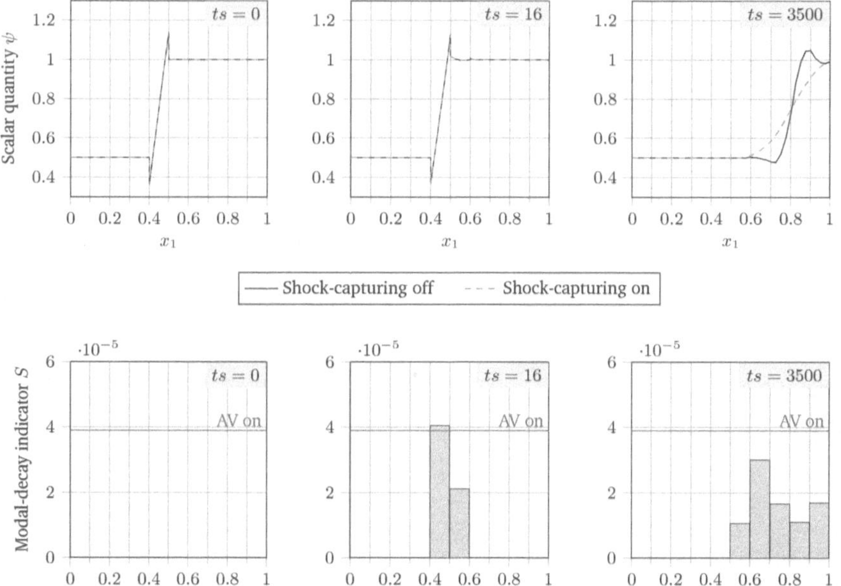

Figure 5.2: Two-step shock-capturing strategy. The modal-decay indicator (5.6) in combination with a piece-wise constant artificial viscosity (AV) (5.31) for the one-dimensional scalar transport equation (5.29). As soon as the solution has been sufficiently smoothed, it stays constant for all time-steps ts. A downstream pollution can be recognized in cells where artificial viscosity has been active.

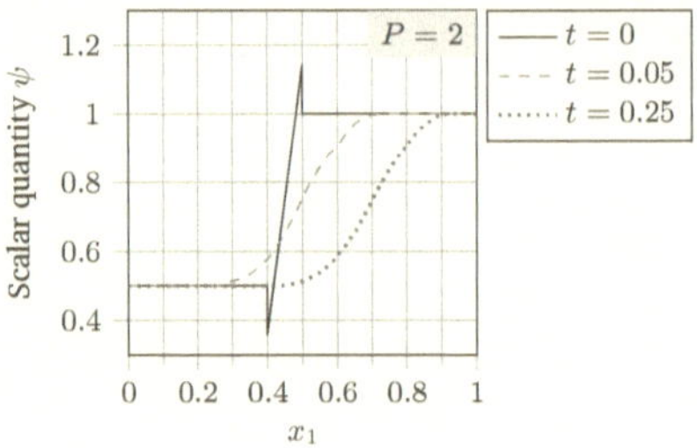

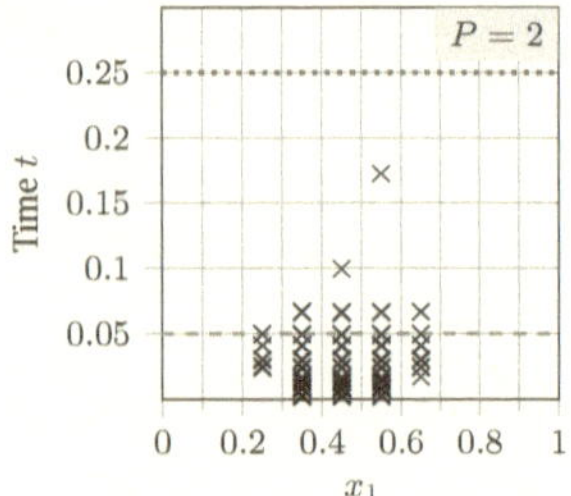

(a) Evolution of the initial jump profile (5.30).

(b) Time-history of cells with active artificial viscosity.

Figure 5.3: Numerical solution and temporal tracking of cells with active artificial viscosity (marked with a cross) for the one-dimensional scalar transport equation (5.29).

In general, the values of S_0 and ε_0 have to be tuned for each test case. For that, a low-resolution run can be used in order to save computational costs. Once the values have been determined, this setting can be applied to high-resolution runs for multiple grid sizes h and polynomial degrees P.

It takes several time-steps until the oscillations reach a certain magnitude and are detected by the indicator. Once the indicator value exceeds the user-defined activation limit, artificial viscosity is added locally. This happens multiple times until the oscillations are damped to a certain level so that the indicator values will always be below the activation limit S_0. High frequencies are still present in the cells through which the shock has been propagating, leading to a downstream pollution of the computational domain.

Figure 5.3a shows the temporal evolution of the initial discontinuity (5.30) at several times $t = \{0, 0.05, 0.25\}$. Until $t \approx 0.07$, the indicator marks cells containing a shock and activates artificial viscosity as shown in Figure 5.3b. Later on, no additional smoothing is applied when neglecting the two outliers. Once the shock has been smoothed to a certain level, it will never return to its former shape in the case of scalar transport.

Figure 5.4 shows refinement studies for grid sizes of $h = \{0.1, 0.05, 0.025\}$ and for polynomial degrees of $P = \{2, 4, 10\}$. The shock becomes steeper for a larger h/P-resolution. For $P = 10$, we obtain sub-cell resolution, meaning that the shock can be resolved within one single cell (Persson and Peraire, 2006; Barter and Darmofal, 2010; Lv et al., 2016).

Inviscid Burgers' equation   In the same manner like for the scalar transport equation (5.29), we analyze the one-dimensional nonlinear inviscid Burgers' equation

$$\frac{\partial \psi(x_1, t)}{\partial t} + \psi(x_1, t)\frac{\partial \psi(x_1, t)}{\partial x_1} = 0\,, \tag{5.34a}$$

$$\psi_0 = \psi(x_1, 0) = \frac{1}{2} + \sin\left(2\pi x_1\right)\,. \tag{5.34b}$$

We choose a periodic computational domain $\Omega = [0, 1]$ with a characteristic grid size of $h = 0.1$. The actual computations are performed in a pseudo-two-dimensional setting. Like in the scalar

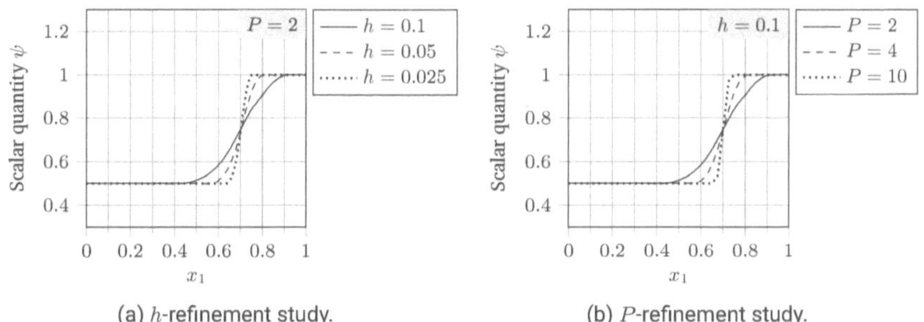

(a) h-refinement study.   (b) P-refinement study.

Figure 5.4: Refinement studies for the one-dimensional scalar transport equation (5.29) using the two-step shock-capturing strategy. The shock thickness scales with $\delta_s \sim h/P$. The shock is contained within a single cell for a large enough h/P-resolution, which is known as *sub-cell resolution*. All curves are plotted at $t = 0.25$.

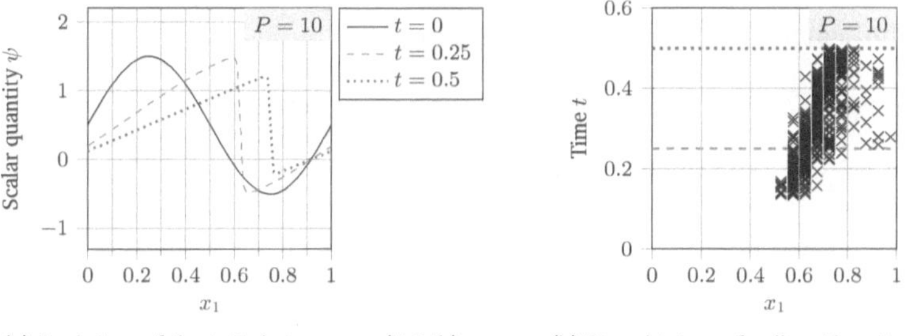

(a) Evolution of the initial sine wave (5.34b).   (b) Time-history of cells with active
artificial viscosity.

Figure 5.5: Numerical solution and temporal tracking of cells with active artificial viscosity (marked with a cross) for the one-dimensional inviscid Burgers' equation (5.34).

transport case, the shock-capturing strategy consists of the modal-decay indicator (5.6) and a simple artificial viscosity on/off-switch (5.31). The on/off-limit is set to $S_0 = 6.25 \cdot 10^{-4}/P^4$ and the maximum artificial viscosity is set to $\varepsilon_0 = 1.0\,h/P$.

Figure 5.5a shows the temporal evolution of the sine wave (5.34b) for a polynomial degree of $P = 10$ for the points in time $t = \{0, 0.25, 0.5\}$. Since the Burgers' equation is self-steepening, the moving front of the sine wave transforms from its initial shape into a shock front. Then, artificial viscosity is activated by the shock indicator and the shock front is flattened. Next, the shock will steepen again and the procedure is repeated. The height of the shock decreases with every diffusion step. This behavior could be verified for low and high polynomial degrees. For $P = 10$, the shock front can be resolved within a single cell, obtaining sub-cell resolution, see again Figure 5.5a. It can clearly be seen that the indicator is still active in cells through which the shock front has been propagating, see Figure 5.5b. The results match well with the results of Persson and Peraire (2006, Figure 1).

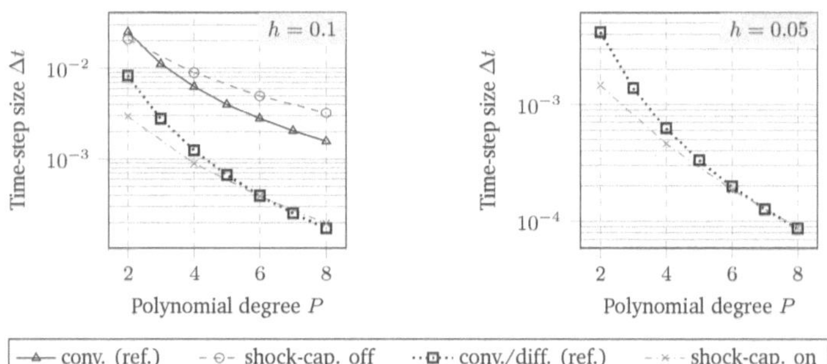

Figure 5.6: Maximum stable time-step sizes using the two-step shock-capturing strategy for the one-dimensional scalar transport equation (5.29). The numerically determined results are compared to a basic convection-diffusion time-step restriction, which serves as a reference, see Equation (5.35).

5.4.2 Influence on the Time-Step Restriction

In order to obtain a numerically stable explicit time-integration scheme for solely convection dominated problems, the CFL-restriction (4.19) has to be fulfilled. An introduction to convective and diffusive time-step restrictions in the context of the DG method has been given in Section 4.1.3.

According to Hesthaven and Warburton (2007, Section 7.5.2), a typical explicit time-step scaling for a convection-diffusion equation reads

$$\Delta t \sim \frac{1}{\underbrace{\lambda_{c,\max}\dfrac{P^2}{h}}_{\text{convective time-scale}} + \underbrace{\varepsilon_0\dfrac{P^4}{h^2}}_{\text{diffusive time-scale}}}. \tag{5.35}$$

In this section, we use this basic variant in order to investigate the influence of the two-step shock-capturing strategy on the maximum admissible time-step size. Subsequently, the maximum stable time-step sizes for the one-dimensional scalar transport equation (5.29) are investigated for test runs with deactivated and activated shock-capturing with $\varepsilon_0 = 10$. The stable time-step sizes are determined by numerical testing. For that, we apply a bisection method with a maximum of 10 iterations in the ranges of $10^{-3} \leq \Delta t \leq 10^{-1}$ and $10^{-5} \leq \Delta t \leq 10^{-2}$ for shock-capturing being activated and deactivated, respectively. The simulation is said to be stable if the discrete version of the energy estimate (3.24)

$$\|\psi_h(\mathbf{x}, t_{n+1})\|_2 \leq \|\psi_h(\mathbf{x}, t_n)\|_2, \qquad \mathbf{x} \in \Omega, \tag{5.36}$$

does not increase during a time span of $t = 0.5$ (Di Pietro and Ern, 2012).

The maximum stable time-step sizes are shown in Figure 5.6, where we use a RK3-method for time-integration, see Figure 4.1. We plot the convective and the convective-diffusive time-step restrictions according to Equation (5.35) as a reference, as well as the numerically determined results on two grids $h = \{0.1, 0.05\}$. The simulations with deactivated shock-capturing being solely convective show a slightly diverging trend for larger values of P compared to the reference. By contrast, the simulations with activated shock-capturing converge to the reference for larger values of P. These investigations indicate that the use of a time-step restriction in the form of Equation (5.35) is suitable for the presented two-step shock-capturing strategy. In the context of the Euler equations (2.5), we use the advanced time-step restrictions (4.24) and (4.25), presented in Section 4.1.3. Furthermore, Figure 5.6 clearly shows that the additional second-order diffusive term decreases the maximum admissible time-step size by an order of magnitude. This shortcoming can be solved by means of an adaptive local time-stepping scheme, see Chapter 4, where the cells are grouped into cell clusters according to their local time-step size. All cells in each cluster are updated with the maximum admissible time-step size of the respective cluster. This saves computational costs, since the time-step restriction is usually only large in cells with artificial viscosity.

5.5 Numerical Results[12]

In this section, we present numerical results of several (pseudo-)two-dimensional test cases for high Mach number flows, see also Figure 5.1 for an overview. We use the Euler equations (5.24) in combination with the two-step shock-capturing strategy presented in Section 5.4. We discretize the convective fluxes and the artificial viscosity flux by a Harten-Lax-van Leer-Compact (HLLC) and an SIP formulation, respectively, see Section 3.2.3. For time-integration, we use standard explicit time-stepping schemes and the adaptive LTS algorithm, see Chapter 4. Subsequently, we combine our approaches with a cut-cell DG IBM, additionally applying a cell-agglomeration technique for small and ill-shaped cut-cells, see Sections 3.3 and 3.3.2.

We select the numerical test cases according to their increasing complexity and methodology:

(i) *Shock-capturing + adaptive LTS*: We investigate the Sod shock tube problem (Sod, 1978) in Section 5.5.1 and a shock-vortex interaction in Section 5.5.2. These first test cases confirm the successful combination of the two-step shock-capturing strategy and the adaptive LTS scheme. Furthermore, the computational costs are analyzed in terms of savings through the adaptive LTS scheme and a brief practical run-time analysis for the Sod shock tube problem.

(ii) *DG IBM + shock-capturing + adaptive LTS*: The results of the novel extension to a cut-cell DG IBM featuring a cell-agglomeration technique is presented for a Sod shock tube variant with an immersed boundary in Section 5.5.3 and a DMR test case in Section 5.5.4. The results are analyzed in the same manner like in part (i).

As we focus on the robustness, the stability, and the applicability in a DG IBM, we do not compare our results to the works of other groups in terms of solution accuracy. The feasibility of the underlying two-step shock-capturing strategy (Persson and Peraire, 2006; Persson, 2013) or similar variants has been successfully demonstrated by various other works (Nguyen et al., 2007; Klöckner et al., 2011; Huerta et al., 2012) in the past years. For more details, the reader is referred to the work by Lv et al. (2016), who presented an extensive study on the performance and accuracy of different shock indicators.

---

[12]Modified version of Geisenhofer et al. (2019, Section 6).

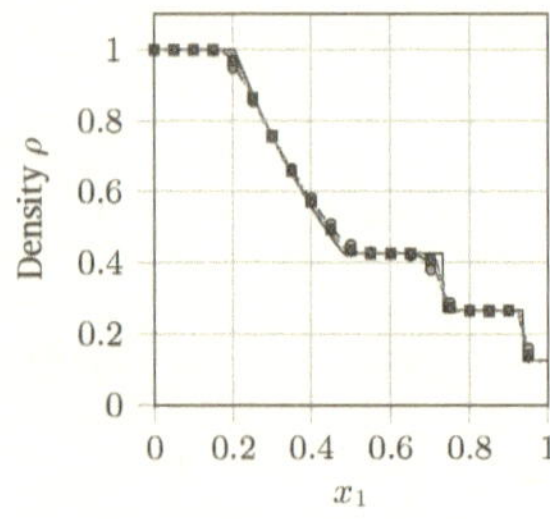
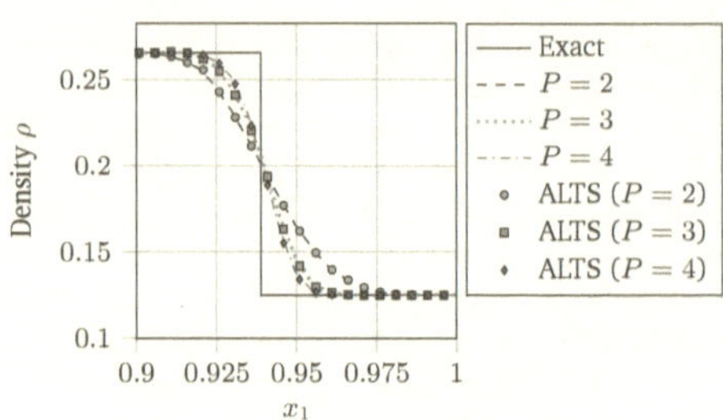

(a) Entire domain, line-out along $x_2 = 0.5$.

(b) Zoom-in around the shock wave.

Figure 5.7: Sod shock tube problem. P-refinement study using the two-step shock-capturing strategy. We show the solution for a global Adams-Bashforth scheme of order $Q = 3$ at $t_{\text{end}} = 0.25$. Additionally, we present comparison results for the adaptive local time-stepping (ALTS) scheme of the same temporal order (adapted from Geisenhofer et al., 2019, Figure 5).

5.5.1 Sod Shock Tube

This classical test case is often considered for the evaluation of numerical schemes regarding their ability to capture simple, non-interacting waves (Sod, 1978). A general introduction to the shock tube problem and its analytical solution procedure has been given in Section 2.2.3.

The initial conditions for the density, the velocity, and the pressure of the variant by Sod (1978) are given by

$$(\rho,\, u_1,\, u_2,\, p)^{\mathsf{T}} = \begin{cases} (1,\, 0,\, 0,\, 1)^{\mathsf{T}}, & \text{for } x_1 \leq 0.5\,, \\ (0.125,\, 0,\, 0,\, 0.1)^{\mathsf{T}}, & \text{for } x_1 > 0.5\,. \end{cases} \tag{5.37}$$

We use a (pseudo-)two-dimensional domain $[0, 1] \times [0, 1]$ with a coarse grid consisting of 50×50 cells with $h = 0.02$. The initial discontinuity, which corresponds to the diaphragm in the experiment, is located at $x_1 = 0.5$. The simulation end time is set to $t_{\text{end}} = 0.25$. Dirichlet boundary conditions are imposed at the left and right boundary of the domain, while adiabatic slip-wall boundary conditions are used at the top and bottom. We compare our numerical results with the exact solution of the one-dimensional Riemann problem as derived in Equations (2.44), (2.46), and (2.48). The shock-capturing parameters are set to $S_0 = 1.0 \cdot 10^{-3}$, $\kappa = 0.5$, and $\varepsilon_0 = 1.0$, see also Section 5.3.1. They are determined for a single polynomial degree and applied for all computations. This verifies the proposed scaling of the indicator value $S_0 \sim 1/P^4$ and the artificial viscosity parameter $\varepsilon \sim h/P$.

Figure 5.7 shows the density profiles for several polynomial degrees $P = \{2, 3, 4\}$. The results are extracted along a horizontal line at $x_2 = 0.5$. For increasing polynomial degrees, the solution converges to the exact solution, capturing the position and the shape of the different waves well. The P-refinement study is conducted with a global Adams-Bashforth (AB) scheme of order $Q = 3$. Additionally, we show the results for the adaptive LTS algorithm of the same order, where we prescribe an initial number of clusters of $C_{\text{init}} = 3$ and set the reclustering interval to $I_{\text{LTS}} = 1$. The adaptive LTS computations match the solution of the global time-stepping scheme without a loss of accuracy.

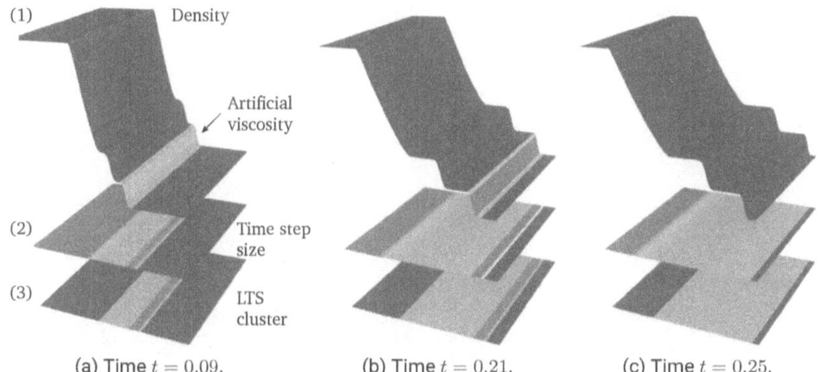

(a) Time $t = 0.09$. (b) Time $t = 0.21$. (c) Time $t = 0.25$.

Figure 5.8: Sod shock tube problem. Illustration of the adaptive LTS algorithm and the two-step shock-capturing strategy. We consider three elevated planes: (1) density profile colored with artificial viscosity, (2) local time-step size, blue/green/red corresponds to large/mid-large/small time-step sizes, (3) LTS clustering, blue/green/red corresponds to small/mid-large/large cell cluster numbers (adapted from Geisenhofer et al., 2019, Figure 6).

An illustration of the temporal evolution is shown in Figure 5.8. The top plane (1) shows the density profile colored with artificial viscosity, the middle plane (2) shows the local time-step size, and the bottom plane (3) depicts the adaptive LTS clustering. The snapshots are taken at three different points in time $t = \{0.09, 0.21, 0.25\}$. It is apparent that the artificial viscosity is localized around the shock wave, since it is the strongest discontinuous feature, see Figures 5.8a and 5.8b. Thus, the time-step sizes are very small in the vicinity of the shock due to the diffusive time-step restriction (4.25). In Figures 5.8a and 5.8b, the LTS clustering consists of three clusters: the untouched area (blue), the segment which is solely dominated by the convective time-step restriction (green), and the segment where artificial viscosity is activated (red). In particular, we choose Figure 5.8c in order to show that there can be several time-steps where the solution is sufficiently smooth and hence artificial viscosity is not added. This yields a more homogeneous distribution of the local time-step sizes, which is why the number of clusters reduces to two.

Computational savings by the adaptive local time-stepping scheme We choose the total number of cell updates N_{tot} as a measure in order to compare the computational costs between a global AB scheme, specified by a reclustering interval of $I_{\text{LTS}} = 0$, and the adaptive LTS scheme. This measure is independent of the actual implementation (quality) and is therefore chosen as a generic benchmark.

In Figure 5.9, we investigate the influence of the reclustering interval I_{LTS} on the total number of cell updates N_{tot} and on the L^2-error $\|\rho_h - \rho_{ex}\|_2$ with respect to the exact solution. A saving of 70.3 % in terms of total cell updates compared to the global AB reference is obtained for $I_{\text{LTS}} = 1$, while maintaining solution accuracy. Additionally, the solution accuracy stays constant for all depicted values of I_{LTS}. It is reasonable that the number of cell updates increases for larger reclustering intervals. This is due to the fact that the clustering has been remaining

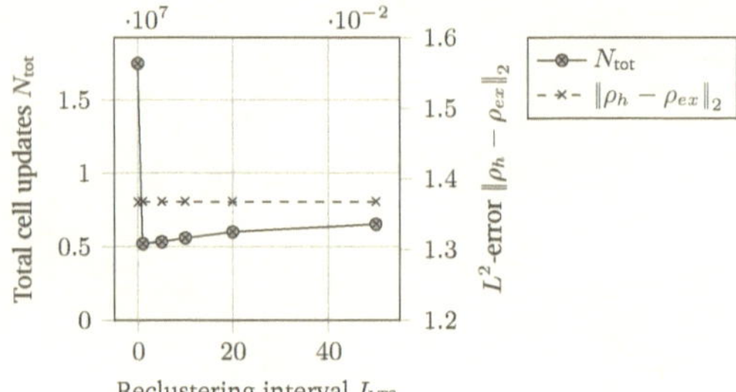

I_{LTS}	N_{tot}	Savings
0	$1.75 \cdot 10^7$	Ref.
1	$0.52 \cdot 10^7$	70.3 %
5	$0.53 \cdot 10^7$	69.7 %
10	$0.56 \cdot 10^7$	68.0 %
20	$0.60 \cdot 10^7$	65.7 %
50	$0.65 \cdot 10^7$	62.9 %

Figure 5.9: Sod shock tube problem. Comparison between a global Adams-Bashforth scheme denoted by $I_{\text{LTS}} = 0$ and the adaptive LTS scheme for several reclustering intervals I_{LTS} in terms of the total number of cell updates N_{tot} and the solution accuracy $\|\rho_h - \rho_{ex}\|_2$. The L^2-error does not change for larger reclustering intervals, whereas the total number of cell updates increases. This is because the clustering does not relay on the physical changes in the flow (adapted from Geisenhofer et al., 2019, Figure 7).

Time-stepping scheme	Run-time	Fraction (run-time)	N_{tot}	Fraction (N_{tot})
GTS	10.4 s	100.0 %	$2.13 \cdot 10^5$	100.0 %
GTS with ALTS code	12.1 s	-16.3 %	$2.13 \cdot 10^5$	100.0 %
ALTS	5.6 s	46.2 %	$0.58 \cdot 10^5$	72.8 %

Table 5.1: Sod shock tube problem. Run-time cost breakdown and comparison between a global time-stepping (GTS) (Adams-Bashforth scheme) and the adaptive local time-stepping (ALTS) scheme. The savings are listed in terms of run-time and the total number of cell updates N_{tot} (adapted from Geisenhofer et al., 2019, Table 1).

in a static state for a longer period of time and has not been updated according to the physical changes in the flow. This forces the time-step size Δt_1 of the slowest cluster to become smaller than necessary, see also Equation (4.31). In practice, the reclustering interval has to be chosen as a compromise between the total number of cell updates and the computational overhead which is produced by the reclustering procedure itself.

Practical savings   In order to give the reader a sense for the savings from a practical point of view, we consider the setting in Table 5.1. We would like to stress that this study is strongly influenced by the implementation quality and the underlying setting. Thus, it is not generalizable. We compare the LTS simulation with a global time-stepping (GTS) simulation. We execute two runs with a standard AB scheme of order $Q = 1$ and the LTS scheme having only one cluster. In this case, both schemes equal the explicit Euler scheme (4.7), but use a different implementation. For the LTS run, we enforce a reclustering in every time-step by setting $I_{\text{LTS}} = 1$ in order to investigate the computationally most expensive scenario. The computational overhead of the reclustering procedure is measured with around 8 % within the LTS run (not shown in Table 5.1). The maximum saving of the adaptive LTS scheme ranges from 46.2 % − 72.8 % for this specific setting, see again Table 5.1. The results are different when using other common explicit time-integration schemes, such as Runge-Kutta (RK) schemes. We abstain from going into the details, since this issue is beyond the scope of this work.

5.5.2 Shock-Vortex Interaction

A common two-dimensional test case for supersonic compressible flow is the interaction of a vortex with a stationary shock wave, which also serves as the title image of this work. A detailed description of the test case can be found in the works by Rault et al. (2003), Dumbser et al. (2014), and the HiOCFD5-Workshop (2017). This test case consists of a complex flow interaction featuring smooth weak and strong discontinuous phenomena and is therefore a challenging test case for high-order numerical schemes.

The computational domain is chosen to be $\Omega = [0, 2] \times [0, 1]$. A stationary normal shock wave with a Mach number of M_s is located at $x_1 = 0.5$. A typical range for the shock Mach number is $M_s \in [1.1, 2]$. The values for density, velocity, and pressure in the pre-shock region are given by

$$(\rho_{\text{pre}}, u_{1,\text{pre}}, u_{2,\text{pre}}, p_{\text{pre}})^\mathsf{T} = \left(1, \sqrt{\gamma \frac{p_{\text{pre}}}{\rho_{\text{pre}}}} M_s, 0, 1\right)^\mathsf{T}, \tag{5.38}$$

where the heat capacity ratio is set to $\gamma = 1.4$. The quantities in the post-shock region can be calculated using the normal shock relations (2.30), resulting in

$$\rho_{\text{post}} = \frac{(\gamma + 1)M_s^2}{2 + (\gamma - 1)M_s^2} \, \rho_{\text{pre}}, \tag{5.39a}$$

$$u_{1,\text{post}} = \frac{2 + (\gamma - 1)M_s^2}{(\gamma + 1)M_s^2} \, u_{1,\text{pre}}, \tag{5.39b}$$

$$u_{2,\text{post}} = 0, \tag{5.39c}$$

$$p_{\text{post}} = \left[1 + \frac{2\gamma}{\gamma + 1}(M_s^2 - 1)\right] p_{\text{pre}}. \tag{5.39d}$$

Additionally, a moving vortex with an angular velocity profile of

$$v_\Phi(r) = \begin{cases} \overline{v_\Phi} \, \dfrac{r}{a_\Phi}, & \text{for } r \le a_\Phi, \\[2mm] \overline{v_\Phi} \, \dfrac{a_\Phi}{a_\Phi{}^2 - b_\Phi{}^2}\left(r - \dfrac{b_\Phi{}^2}{r}\right), & \text{for } a_\Phi < r \le b_\Phi, \\[2mm] 0, & \text{otherwise}, \end{cases} \tag{5.40}$$

is superimposed onto the initial conditions (5.38). The radius of the vortex is denoted by $r = \sqrt{(x_1 - x_1^c)^2 + (x_2 - x_2^c)^2}$ and the mean angular velocity is $\overline{v_\Phi}$. The initial center of the vortex is located at $(x_{1,0}, x_{2,0})^\mathsf{T} = (0.25, 0.5)^\mathsf{T}$. The initial temperature distribution of the vortex can be obtained by solving the ordinary differential equation (ODE)

$$\frac{\mathrm{d}T}{\mathrm{d}r} = \frac{\gamma - 1}{R_{\text{gas}}\gamma} \frac{v_\Phi{}^2(r)}{r}. \tag{5.41}$$

We calculate the temperature profile $T(r)$ by integrating Equation (5.41) with respect to r, resulting in

$$T(r) = \frac{\gamma - 1}{R_{\text{gas}}\gamma} \begin{cases} T(a_\Phi) - \dfrac{\gamma-1}{R_{\text{gas}}\gamma} \dfrac{\overline{v_\Phi}^2}{2}\left(1 - \dfrac{r^2}{a_\Phi{}^2}\right), & \text{for } r \le a_\Phi, \\[3mm] T(b_\Phi) - \dfrac{\gamma-1}{R_{\text{gas}}\gamma}\left(\dfrac{\overline{v_\Phi} a_\Phi}{a_\Phi{}^2 - b_\Phi{}^2}\right)^2 \\[2mm] \quad \left(-\dfrac{r^2}{2} + 2b_\Phi{}^2 \ln(r) + \dfrac{b_\Phi{}^4}{2r^2} - 2b_\Phi{}^2 \ln(b_\Phi)\right), & \text{for } a_\Phi < r \le b_\Phi, \\[3mm] 0, & \text{otherwise}. \end{cases} \tag{5.42}$$

The integration constants $T(a_\Phi)$ and $T(b_\Phi)$ are obtained by $T(a_\Phi) = T(a_\Phi)_{a_\Phi \leq r \leq b_\Phi}$ for $r \leq a_\Phi$ and $T(b_\Phi) = T_{\mathrm{pre}}$ for $a_\Phi \leq r \leq b_\Phi$, respectively. The density and pressure distributions are calculated by the isentropic relations

$$\rho = \rho_\infty \left(\frac{T}{T_\infty} \right)^{\frac{1}{\gamma-1}}, \qquad p = p_\infty \left(\frac{T}{T_\infty} \right)^{\frac{\gamma}{\gamma-1}}. \tag{5.43}$$

The reference values are based on the undisturbed flow values upstream of the shock wave, yielding $p_\infty = R_{\mathrm{gas}}\rho_\infty T_\infty$, where the gas constant is set to $R_{\mathrm{gas}} = 1$ for the sake of simplicity. Typically, the strength of the vortex is varied by its Mach number $\mathrm{M_v} = \overline{v_\Phi}/a_\infty$, where $a_\infty = \sqrt{\gamma p_\infty/\rho_\infty}$ represents the upstream speed of sound. We set further parameters to $a_\Phi = 0.075$, $b_\Phi = 0.175$, $\mathrm{M_v} = 0.9$, and $\mathrm{M_s} = 1.5$. The values $\rho_\infty = \rho_{\mathrm{pre}} = 1$ and $p_\infty = p_{\mathrm{pre}} = 1$ are given by the undisturbed upstream conditions. A typical range for the simulation end time is $t_{\mathrm{end}} \in [0.7, 0.8]$.

Physics behind shock-vortex interactions   For a deeper insight into the physical phenomenology of this test case, we recommend the work by Rault et al. (2003). In the following, we summarize the main aspects of this work and give a brief introduction to the topic.

Interactions of shock waves and vortex-like structures occur in multiple practical applications such as in supersonic aircraft. Here, vortices are created at the front of the airplane and are convected downstream, where they may interact with shock waves generated by the airplane itself or the engine intake (Smart et al., 1998). Shock-generated noise is also an active field of research in today's jet engine development. In general, the flow field of a shock-vortex interaction can become very complex, depending on the structure and strength of both the vortex $\mathrm{M_v}$ and the shock wave $\mathrm{M_s}$. If the vortex is weak relative to the shock, the shock is only slightly distorted by the vortex so that a theoretical analysis which assumes only small perturbations of the planar shock front can be applied (Ribner, 1985). Interactions of strong vortices lead to a distortion of the shock in an S-like structure, see also the Schlieren distribution at $t = 0.16$ in Figure 5.10. This can lead to two diffracted shocks which are connected by a refracted shock passing through the core of the vortex (Ellzey et al., 1995; Chatterjee, 1999), if the vortex is stronger than the value of $\mathrm{M_v} = 0.9$ chosen in this work. For very strong shocks of $\mathrm{M_s} > 3.0$, a complex wave pattern establishes, since the vortex is often split into several smaller structures directly after hitting the shock wave. If the shock is weak relative to the vortex, the interaction creates a reflection similar to a regular reflection. When increasing the strength of the shock and keeping the vortex unchanged, a Mach reflection structure develops instead. We refer to Section 5.5.4 for an introduction to shock wave reflections. The setting of this test case with $(\mathrm{M_v}, \mathrm{M_s}) = (0.9, 1.5)$ is called a *strong vortex/strong shock* interaction.

Another aspect of the interaction is the generation and evolution of acoustic waves. Here, mainly the form of the acoustic wave after the interaction was investigated in the literature (Ribner, 1985; Grasso and Pirozzoli, 2000). For weak and strong vortices, a cylindrical acoustic wave is generated, which later propagates radially outwards from the vortex. For weak interactions, the linear theory of sound production can be applied, whereas for strong interactions, the complexity limits the application of linear theory.

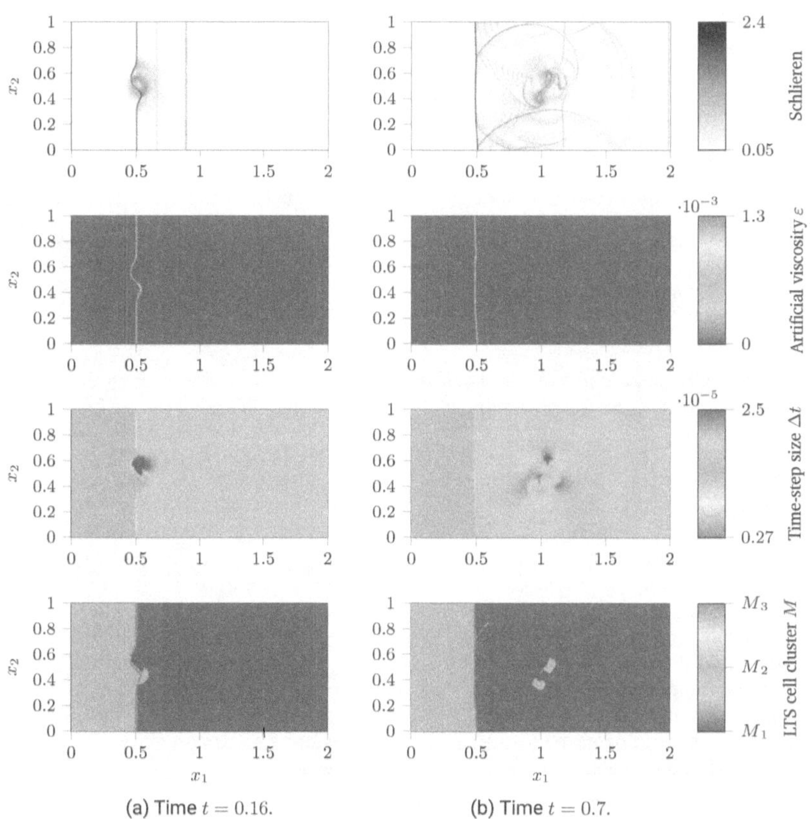

Figure 5.10: Shock-vortex interaction. The results are computed for a vortex Mach number of $M_v = 0.9$ and a shock Mach number of $M_s = 1.5$. We show the Schlieren distribution calculated by $\ln(1 + |\nabla\rho|_2)/\ln(10)$ (first row), the artificial viscosity distribution (second row), the time-step sizes (third row), and the LTS cell clustering (last row) using the presented adaptive LTS scheme at $t = 0.16$ (left column) and $t = 0.7$ (right column). This test case serves as the title image of this work (adapted from Geisenhofer et al., 2019, Figure 8).

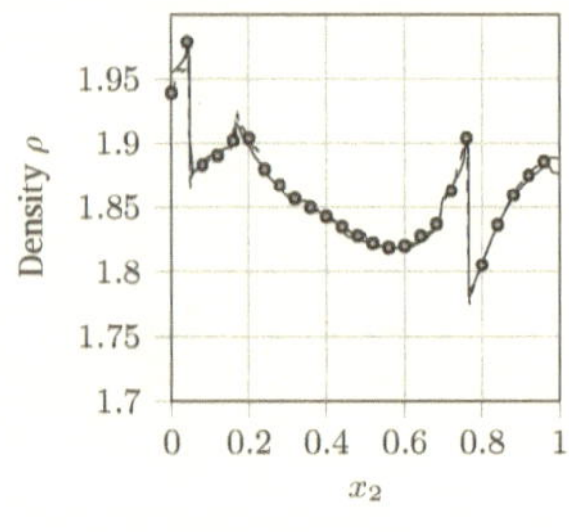
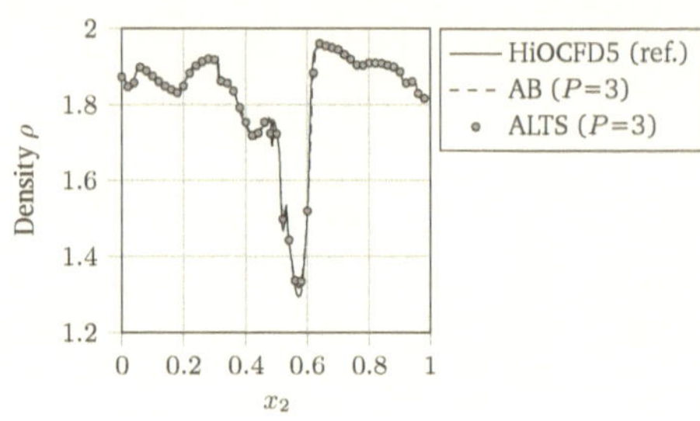

(a) Vertical line at $x_1 = 0.52 + \varepsilon_x$.   (b) Vertical line at $x_1 = 1.05 + \varepsilon_x$.

Figure 5.11: Shock-vortex interaction. We show the density distribution along two vertical lines shifted with $\varepsilon_x = 1.0 \cdot 10^{-4}$ at $t = 0.7$. Line (a) is placed directly behind the shock in the post-shock region. Line (b) goes through the vortex structure. We apply the two-step shock-capturing strategy in combination with the adaptive local time-stepping (ALTS) scheme and a global Adams-Bashforth scheme, both of order $Q = 3$, for a comparison. The results are in good agreement with the literature (adapted from Geisenhofer et al., 2019, Figure 9).

Video   We uploaded a video of the shock-vortex interaction in order to show the temporal development of the structure (click here)[13]. The simulation parameters can be found online.

Results   In Figure 5.10, we report the results on a computational grid consisting of 600×300 cells with $h = 0.00\overline{3}$. We show the pressure and artificial viscosity distribution for two different times $t = \{0.16, 0.7\}$ in the first and second column, respectively. We present the time-step sizes in the third row and the respective LTS clustering in the last row for a polynomial degree of $P = 3$. Further simulation parameters are $Q = 3$, $I_{\mathrm{LTS}} = 1$, and $C_{\mathrm{init}} = 3$. At $t = 0.16$, the shock is distorted by the moving vortex. The distortion strongly depends on the strength of the vortex compared to the strength of the shock (Chatterjee, 1999). The interaction between the vortex and the shock produces sound waves which propagate downstream. The initial configuration creates a right running wave that is similar to the numerical artifacts encountered in the DMR test case (Woodward and Colella, 1984), see Section 5.5.4. The vast majority of artificial viscosity is added around the shock, which results in an almost static LTS clustering except for the outliers caused by the generated sound waves.

In Figure 5.11, we show the density distribution along two vertical lines $x_1 = 0.52 + \varepsilon_x$ and $x_1 = 1.05 + \varepsilon_x$ for a quantitative assessment of the solution accuracy. The vertical lines are slightly shifted by a small parameter $\varepsilon_x = 1.0 \cdot 10^{-4}$ in order to assess the accuracy of high-order schemes properly by not considering the solution on cell boundaries (HiOCFD5-Workshop, 2017). The reference values are extracted from a contribution to the HiOCFD5-Workshop (2017) by You and Kim (2017) based on a multi-dimensional limiting strategy (Park and Kim, 2014) on a mixed rectangular-triangular grid with $h = 1/250$ and a polynomial degree of $P = 3$. Our results are in very good agreement with the reference. Furthermore, we perform a simulation for a global AB scheme of order $Q = 3$, which leads to a total number of $N_{\mathrm{tot}} \approx 3.19 \cdot 10^{10}$ cell updates. The adaptive LTS simulation only performs $N_{\mathrm{tot}} \approx 0.98 \cdot 10^{10}$ updates, which results in a saving of approximately 69 %.

---

[13]Video of a shock-vortex interaction, https://doi.org/10.25534/tudatalib-307, visited on 10/07/2020.

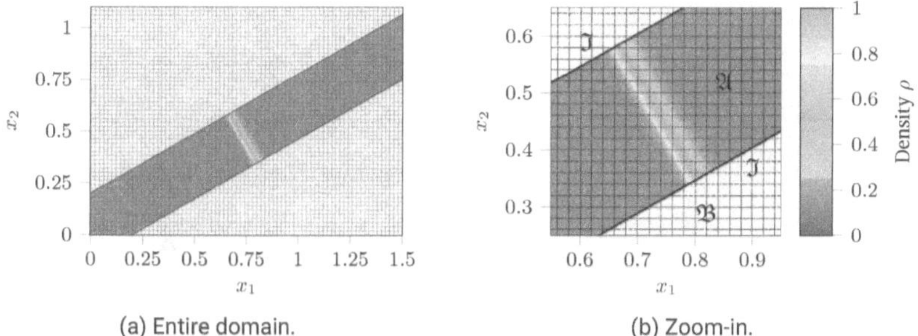

(a) Entire domain.   (b) Zoom-in.

Figure 5.12: Sod shock tube computed with a DG IBM. We show the smoothed initial density distribution (5.37) on a Cartesian grid, where the fluid part $\mathfrak{A}$ is cut out by the zero iso-contour of a level-set function. The void part is denoted by $\mathfrak{B}$ and the interface by $\mathfrak{J}$ (adapted from Geisenhofer et al., 2019, Figure 10).

5.5.3 IBM Sod Shock Tube

We reuse the Sod shock tube problem (Sod, 1978) presented in Section 5.5.1 in order to show the viability of our novel combination of a cut-cell DG IBM, a two-step shock-capturing strategy and an adaptive LTS scheme, as presented in the beginning of this chapter. We set the level-set function to

$$\varphi(\mathbf{x}) = - \left[x_2 - \tan(\pi/6)(x_1 - 0.2) \right] \left[x_2 - \tan(\pi/6)(x_1 + 0.2) \right] , \qquad (5.44)$$

in order to create a configuration where we enforce cell-agglomeration by choosing a threshold of $\delta_{\mathrm{agg}} = 0.3$. The computational domain is $\Omega = [0, 1.5] \times [0, 1.1]$ with a grid consisting of 75×55 cells with $h = 0.02$. The interface $\mathfrak{J}$ cuts the grid as depicted in Figure 5.12, splitting the domain into the fluid part $\mathfrak{A}$ and the void part $\mathfrak{B}$. The discontinuous initial conditions (5.37) are smoothed over several cells in order to not introduce an error by an oscillating initial projection. The shock indicator values are calculated using the extended variant (5.8).

Performance of the two-step shock-capturing strategy on cut-cells   In the following, we evaluate the performance of the modal-decay indicator (5.8) on an agglomerated cut-cell grid $\mathcal{K}_h^{\mathrm{X,agg}}$ in order to show its successful extension to a DG IBM. We reuse the shock-capturing parameters $S_0 = 1.0 \cdot 10^{-3}$, $\kappa = 0.5$, and $\varepsilon_0 = 1.0$ from the boundary-fitted case in Section 5.5.1 without any modifications. We use the adaptive LTS scheme with the parameters $Q = 3$, $I_{\mathrm{LTS}} = 1$, and $C_{\mathrm{init}} = 2$. We perform h-/P-refinement studies, in order to verify the scaling of the two-step shock-capturing strategy in a DG IBM. Figure 5.13 shows the density distributions at a zoom-in around the shock wave at $t_{\mathrm{end}} = 0.25$. In both refinement studies, the numerical approximations converge towards the exact solution for an increasing h/P-resolution. These investigations confirm that the proposed $1/P^4$-scaling of the activating indicator value S_0 and the h/P-scaling of the artificial viscosity formula (5.20) are also valid in a DG IBM.

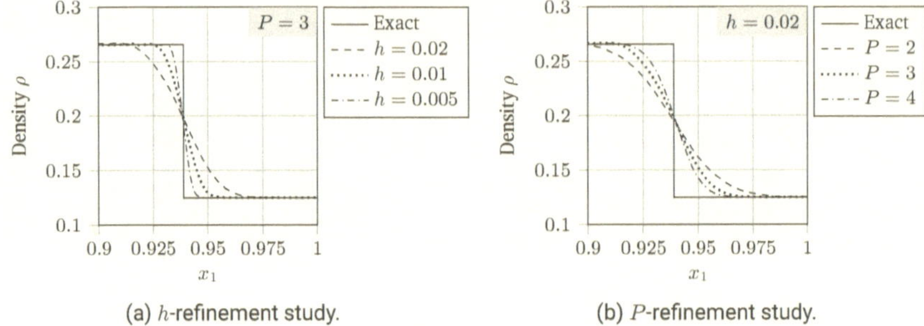

(a) h-refinement study. (b) P-refinement study.

Figure 5.13: Sod shock tube computed with a DG IBM. The refinement studies verify the applicability of the two-step shock-capturing strategy and the adaptive LTS scheme on an agglomerated cut-cell grid $\mathcal{K}_h^{\mathrm{X,agg}}$. The $1/P^4$-scaling of the critical indicator value S_0 which activates artificial viscosity and the h/P-scaling of the artificial viscosity determination formula (5.20) are valid in a DG IBM. The plots show the density distributions at $t_{\mathrm{end}} = 0.25$ (adapted from Geisenhofer et al., 2019, Figure 11).

Computational savings of the adaptive local time-stepping scheme on cut-cells We compare the results of the adaptive LTS scheme in terms of accuracy to the exact solution of the shock tube problem, see also Section 2.2.3. Additionally, we investigate the computational savings measured by the total number of cell updates N_{tot} compared to a global AB time-stepping scheme in the same manner like for the boundary-fitted case in Section 5.5.1. We show the results in Figure 5.14 for a setting with $P = 3$, $Q = 3$, $I_{\mathrm{LTS}} = 1$, and $C_{\mathrm{init}} = 2$. The use of the adaptive LTS scheme results in a saving of around 63 % in terms of the total number of cell updates N_{tot} for different reclustering intervals I_{LTS}. In contrast to the results of the boundary-fitted case shown in Figure 5.9, the savings remain approximately the same for different values of I_{LTS}. This fact strongly depends on the LTS clustering which is dominated by small non-agglomerated cut-cells. In the presented test case, the clustering consists of solely two clusters, compared to three clusters in the boundary-fitted case. In the presented DG IBM, the number of sub-steps between the different LTS clusters varies up to two orders of magnitude. However, we did not experience any inconsistencies caused by this issue during our numerical experiments. As a remedy, the underlying K-means clustering algorithm (Macqueen, 1967) may be tuned in a way so that it produces a clustering with a different number of clusters, which may lead to larger savings. We abstain from performing additional studies, since we focus on the applicability of the method itself.

In Table 5.2, we list the computational savings of the adaptive LTS scheme for the P-refinement study shown in Figure 5.13b. The savings are calculated with respect to a reference simulation with a global AB scheme of the same temporal order $Q = 3$. The results indicate that higher savings in terms of the total number of cell updates N_{tot} can be obtained for higher polynomial degrees P. This is due to the diffusive time-step restriction (4.25), where the time-step size scales with $\Delta t \sim 1/P^2$, while keeping h fixed. In general, this effect strongly depends on the topology of the test case and, thus, on the LTS clustering. However, it is only slightly present in this test case.

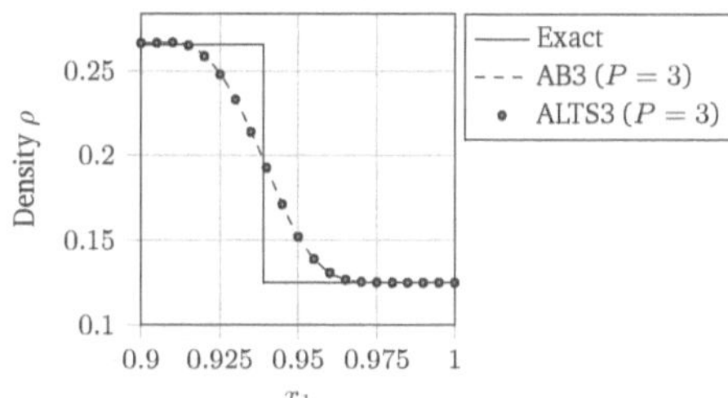

I_{LTS}	N_{tot}	Savings
0	$3.09 \cdot 10^8$	Ref.
1	$1.14 \cdot 10^8$	63.1 %
5	$1.14 \cdot 10^8$	63.1 %
10	$1.14 \cdot 10^8$	63.1 %
20	$1.14 \cdot 10^8$	63.1 %
50	$1.15 \cdot 10^8$	62.8 %

Figure 5.14: Sod shock tube computed with a DG IBM. Comparison between a global Adams-Bashforth scheme ($I_{\mathrm{LTS}} = 0$) and the adaptive local time-stepping (ALTS) scheme for several reclustering intervals I_{LTS} in terms of the total number of cell updates N_{tot} and the density distribution ρ. The corresponding grid is shown in Figure 5.12 (adapted from Geisenhofer et al., 2019, Figure 12).

P	I_{LTS}	N_{tot}	Ref.	Savings
2	1	$0.64 \cdot 10^8$	$1.72 \cdot 10^8$	62.8 %
3	1	$1.15 \cdot 10^8$	$3.09 \cdot 10^8$	62.8 %
4	1	$2.31 \cdot 10^8$	$6.40 \cdot 10^8$	63.9 %

Table 5.2: Sod shock tube computed with a DG IBM. We list the computational savings in terms of the total number of cell updates N_{tot} for a reclustering interval of $I_{\mathrm{LTS}} = 1$ and several polynomial degrees P. We compare the adaptive LTS scheme with a global time-stepping scheme of the same temporal order $Q = 3$ for each polynomial degree (adapted from Geisenhofer et al., 2019, Table 2).

5.5.4 IBM Double Mach Reflection

As a final test case of this chapter, we investigate the complex two-dimensional DMR, which was first proposed by Woodward and Colella (1984) as a numerical benchmark. This self-similar flow configuration contains several challenging features such as two moving triple points as well as complicated shock reflections and interactions. Experimentally, the DMR can be set up by creating a supersonic flow that hits a ramp or a wedge. The reader is referred to Section 2.2.4 for a brief introduction into the theory and concepts of shock wave reflection phenomena.

The underlying setting is based on a shock wave with a Mach number of $M_s = 10$ hitting a reflecting wall which is inclined at an angle of $\pi/6$. We consider an ideal gas with a heat capacity ratio of $\gamma = 1.4$. The quantities on the right side of the incident shock are at rest, whereas on the left side, the corresponding post-shock conditions are prescribed satisfying the normal shock relations (2.30). The initial conditions are given by

$$(\rho, u_1, u_2, p)^{\mathsf{T}} = \begin{cases} (8, 8.25, 0, 116.5)^{\mathsf{T}} & \text{for } x_1 < 0.16\,, \\ (1, 0, 0, 1)^{\mathsf{T}} & \text{for } x_1 \geq 0.16\,. \end{cases} \tag{5.45}$$

The computational domain is chosen as $\Omega = [0,3] \times [0,2]$ together with a grid consisting of 300×200 cells with $h = 0.02$ for the DG IBM case. We assign supersonic boundary conditions to the inlet, the bottom boundary for $x_1 < 0.16$, the top boundary, and the outlet. Additionally, we prescribe the exact movement of the incident shock wave at the top boundary. For $x_1 \geq 0.16$, we apply an adiabatic slip-wall boundary condition along the ramp. If the position $x_1 = 0.16$ where the initial conditions change is not exactly located on a cell boundary, we apply a smoothing over several cells in order to not introduce additional oscillations by the initial projection. We choose an agglomeration threshold of $\delta_{\text{agg}} = 0.3$. The indicator value which activates artificial viscosity is set to $S_0 = 1.0 \cdot 10^{-4}$. Severe oscillations can occur due to the abrupt, nonphysical change of the boundary conditions at $x_1 = 0.16$. While these oscillations can be damped via an excessive amount of artificial viscosity, we decide to set the characteristic velocity to a constant value of $\lambda_{\text{c,max}} = 25$ in order to guarantee a sufficient smoothing during the development phase of the self-similar structure. We apply the adaptive LTS scheme with an order of $Q = 3$, a reclustering interval of $I_{\text{LTS}} = 1$, and an initial number of clusters of $C_{\text{init}} = 2$. Note that the employed K-means clustering algorithm (Macqueen, 1967), see also Section 4.2, reduces the number of LTS clusters to two, even if $C_{\text{init}} > 2$ is prescribed. In this test case, the largest flow velocity $|\mathbf{u}| + a$ given by the pre-shock state with a Mach number of $M_s = 10$ has a large influence on the LTS clustering, since the time-step size of most cells in the fluid domain is limited by the convective time-step restriction

$$\Delta t_c \leq \frac{C_{\text{CFL}}}{2P+1} \frac{h}{|\mathbf{u}| + a}\,. \tag{4.24, repeated}$$

Only a comparably small number of non-agglomerated cut-cells and cells with active artificial viscosity place a larger restriction on the time-step size so that the LTS clustering is dominated by the flow velocity of the pre- and post-shock state, respectively, in this test case. Note that the non-agglomerated cut-cells with active artificial viscosity prescribe the time-step size of LTS cluster M_2 in the pre-shock state, see Figure 5.15. An adaption of the K-means clustering algorithm (Macqueen, 1967) may result in an LTS clustering consisting of three cell clusters, which may lead to lower computational costs.

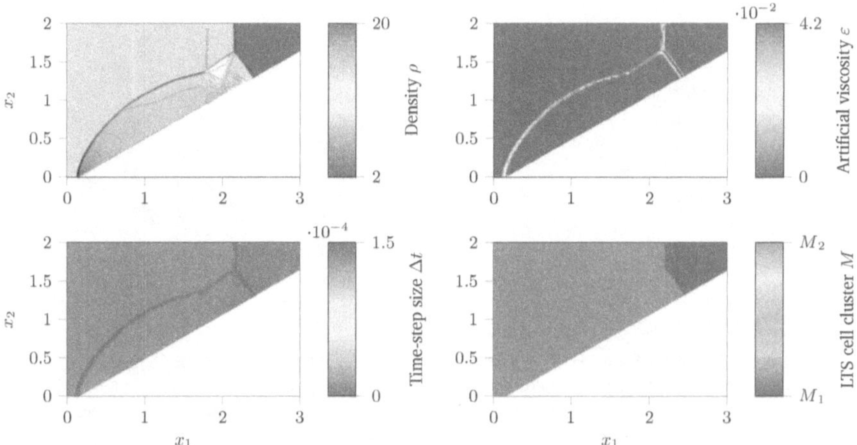

Figure 5.15: Double Mach reflection computed with a DG IBM. The results are obtained for a polynomial degree of $P = 3$, using the presented cell-agglomeration technique with a threshold of $\delta_{\mathrm{agg}} = 0.3$. We show the density distribution with 20 equally distributed contour lines, the artificial viscosity distribution, the time-step sizes, and the cell clustering at $t = 0.2$, applying the adaptive LTS scheme (adapted from Geisenhofer et al., 2019, Figure 13).

Video We uploaded a video of the DMR in order to show the temporal development of the structure (click here)[14]. The simulation parameters can be found online.

Results The results of the DG IBM case are depicted in Figure 5.15. The density and artificial viscosity distributions are shown with the corresponding time-stepping information for a polynomial degree of $P = 3$. Note that the color scale varies from plot to plot for the sake of visibility. The thin vertical stripes, for example, visible at $x_1 \approx 0.9$ and $x_1 \approx 1.8$, are numerical artifacts caused by the initial condition and the top boundary condition, respectively (Woodward and Colella, 1984). There is no general way to prevent the creation of nonphysical waves emanating from a non-grid-aligned oblique shock in the context of solving this oblique Riemann problem (Kemm, 2014). The top boundary condition is dynamically adapted with the exact shock velocity. Additionally, we apply a smoothing over several cells as done for the initial condition. Most troubled cells are located around the incident shock wave, the reflected shock waves, and the Mach stem, which emanates from the first triple point. Challenging parts for numerical schemes are the curly flow structure at the bottom right part of the DMR and the Kelvin-Helmholtz instabilities along the first slip line. In order to resolve these flow phenomena, adjustments of the underlying shock-capturing scheme are necessary, for example, by an appropriate a-posteriori limiting (Dumbser et al., 2014) or a dynamic threshold setting (Lv et al., 2016) if the resolution is not increased to a maximum. We have resolved the instabilities with the presented shock-capturing strategy for a larger end time, which has the same effect as a finer resolution due to the self-similarity of the flow structure, as shown in the video.

---

[14]Video of a DMR, https://doi.org/10.25534/tudatalib-308, visited on 10/07/2020.

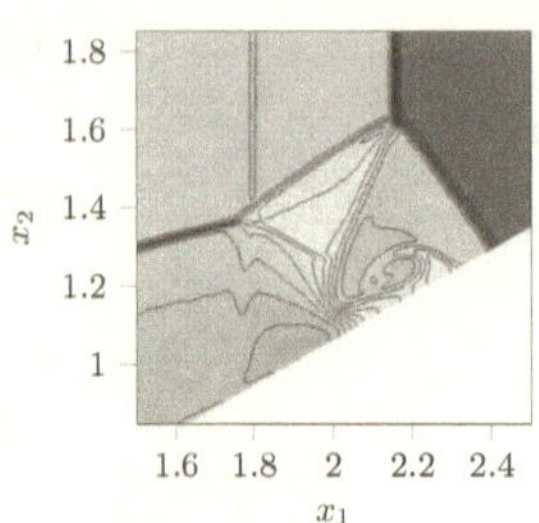
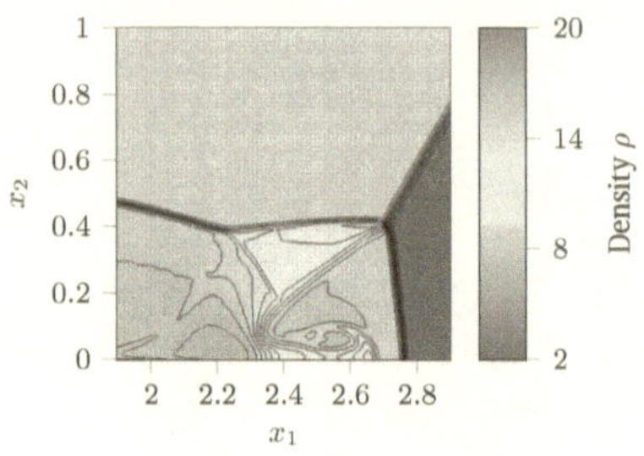

(a) DG immersed boundary method (DG IBM).   (b) Boundary-fitted DG method.

Figure 5.16: Double Mach reflection. Comparison between a DG IBM, see also Figure 5.15, and a boundary-fitted simulation using a global Adams-Bashforth scheme of order $Q = 3$ for a polynomial degree of $P = 3$. A zoom-in around the curly flow structure is plotted with 20 equally distributed contour lines for both cases. (b) The domain size is $\Omega = [0, 4] \times [0, 1]$ with 300×75 cells. Other parameters are chosen like in the DG IBM test case (adapted from Geisenhofer et al., 2019, Figure 14).

Especially in this DG IBM test case with a very strong incident shock wave, the adaptive LTS scheme together with the presented cell-agglomeration technique shows its potential compared to a global time-stepping scheme. This is due to the fact that the time-step sizes in non-agglomerated cut-cells with activated artificial viscosity can differ up to two orders of magnitude compared to non-troubled standard cells in this test case. Suitable parallelization techniques in the context an adaptive LTS scheme start to play an important role for such complex test cases, see Section 4.4 and the work by Weber (2018) for a deeper insight into this topic. A detailed assessment of the computational run-times is not presented for this test case, since it is beyond the scope of this work. The reader is referred to the previous sections for a basic analysis of the computational costs.

For a comparison of the solution accuracy, we additionally present the results on a boundary-fitted grid for a global AB time-stepping scheme in Figure 5.16b. We choose a comparable spatial resolution to the DG IBM test case depicted in Figure 5.16a. The results are in very good agreement, for example, when considering the position of the triple points or the shape of the curly flow structure. In conclusion, it can be stated that the combination of the DG IBM, the two-step shock-capturing strategy, and the adaptive LTS scheme maintains the accuracy of the numerical solution.

5.6 Summary[15]

In this chapter, we have presented a novel efficient combination of a cut-cell DG IBM for high Mach number flows consisting of a two-step shock-capturing strategy based on artificial viscosity and an adaptive explicit LTS scheme. This combination makes use of a cell-agglomeration

---

[15]Extended version of Geisenhofer et al. (2019, Section 7).

technique, avoiding problems with small or ill-shaped cut-cells. The reader is referred to Figure 5.1 for an overview of the performed numerical investigations.

Non-agglomerated cut-cells with activated artificial viscosity represent a worst-case scenario in terms of the maximum admissible time-step size for explicit time-integration schemes, see Equation (4.25). The time-step size of these cells can differ up to two orders of magnitude compared to standard cells on the background grid in the presented test cases. Both the adaptive LTS scheme and the cell-agglomeration technique significantly decrease the otherwise large computational costs, while not introducing additional spatial or temporal errors.

The effectiveness of the novel combination has been verified in three (pseudo-)two-dimensional test cases in terms of robustness, stability, and accuracy. For example, we have compared our numerical results with the exact solution of the Sod shock tube problem for several polynomial degrees $P \geq 2$ in a boundary-fitted and an immersed boundary configuration, see Sections 5.5.1 and 5.5.3. In both cases, the same error levels are obtained for the adaptive LTS scheme like for a global AB scheme. Up to $63 - 70\,\%$ of the number of the total cell updates are saved by using the adaptive LTS scheme depending on the reclustering interval. Furthermore, around $46\,\%$ of the practical run-time are saved for the Sod shock tube problem on a boundary-fitted grid. However, this saving highly depends on the implementation and the problem setting. Furthermore, we have presented numerical results for complex two-dimensional test cases, such as a shock-vortex interaction and a DMR, which shows the geometrical flexibility of the presented DG IBM by representing solids by the zero iso-contour of a level-set function. All results are in good agreement with the literature.

For all test cases, we have applied a two-step shock-capturing based on the works by Persson and Peraire (2006) and Persson (2013) without major modifications except for the extension to a cut-cell grid and the computation of artificial viscosity (Barter and Darmofal, 2007; Klöckner et al., 2011), see Sections 5.2.1 and 5.3.1, respectively. It is essential to project the piece-wise constant artificial viscosity field onto a C^0-continuous field in order to obtain stable and accurate solutions.

In Table A.1, we state the simulation parameters of the test cases presented in Sections 5.4 and 5.5 for the sake of reproducibility.

6 High-Order Shock-Fitting Using an Extended Discontinuous Galerkin Method

This chapter deals with the methodology of shock-fitting and in particular with a novel reconstruction technique of a sharp and sub-cell accurate shock front in the context of an extended discontinuous Galerkin (XDG) method for supersonic compressible flow. The shock front is represented by the zero iso-contour of a level-set function.

Shock-capturing strategies spoil the global high-order accuracy normally found in discontinuous Galerkin (DG) methods, among others, due to the use of artificial viscosity. Furthermore, the loss of global high-order accuracy is shared by all total variation diminishing (TVD) schemes (Shu, 1987). We refer to the work by LeVeque (1999) for a detailed overview of different limiting approaches. As an alternative, total variation bounded (TVB) limiting approaches, such as the *minmod*-type limiter by Shu (1987), try to overcome the local degeneration to first-order accuracy at discontinuities. In general, it is difficult to design a limiter, which achieves both high-order accuracy and a non-oscillatory property in the vicinity of discontinuities (Shu, 2016). Weighted essentially non-oscillatory (WENO) schemes try to combine both desired properties, but usually destroy the locality of the numerical scheme. Additionally, WENO approaches are complicated to implement, especially on unstructured grids and in the higher-dimensional case, since the information of neighbors' neighbors is needed (Qiu and Shu, 2005b; Zhu et al., 2008). Follow-on works try to restore the locality on structured and unstructured grids (Zhong and Shu, 2013; Zhu et al., 2013).

In this work, we aim for a sharp interface description of the shock front in order to regain the high-order accuracy before and after the shock. For this approach, it is essential to develop a methodology which delivers a highly accurate shock front in terms of position and shape before the entire flow field is computed. We focus on steady test cases, where the shock front is fixed in time and space. The extension to unsteady flows with moving discontinuous flow phenomena increases the complexity strongly, since algorithms for the movement of the interface are essentially needed (Utz and Kummer, 2017; Utz, 2018; Kummer et al., 2018). At this point, it is not foreseeable whether additional stabilization mechanisms are required so that we exclude unsteady flow scenarios from this work.

In Sections 6.1 and 6.2, we present the state of the art of shock-fitting approaches. Since the position and the shape of non-simple oblique and curved shock waves can in general not be determined a-priori, we derive a novel reconstruction technique of a shock level-set function based on a standard DG simulation with shock-capturing in Section 6.3. The reconstructed shock front is iteratively adapted by means of an implicit pseudo-time-stepping procedure until it coincides with the exact shock position. We present a one-dimensional proof of concept in Section 6.4. In particular, suitable indicators for the correction of the shock position are derived. We close the chapter with a brief summary in Section 6.5.

6.1 State of the Art[16]

The basic idea of shock-fitting originates in the computation of a supersonic flow over blunt bodies as found in supersonic aircraft or, even more advanced, during the reentry of space shuttles or of their autonomous successor, the *Dream Chaser* (Krevor et al., 2011), into the terrestrial atmosphere. In all cases, a curved and detached shock wave, often called a *bow shock*, forms in an a-priori unknown distance in front of the blunt body. This bow shock raises two difficult challenges. The unknown shock position prohibits the use of static grid-alignment. Additionally, high-order numerical schemes are inherently unstable near discontinuities. Shock-capturing approaches meet both of these challenges by limiting the numerical solution in the vicinity of the discontinuities or by smoothing the numerical solution with an sufficient amount of artificial viscosity. By contrast, classical shock-fitting approaches circumvent these challenges by modifying the computational domain such that a domain boundary coincides with the shock front. In this case, the computational domain only contains the post-shock region, for which the values are prescribed by the shock relations across the shock boundary. This renders shock-fitting approaches attractive for high-order XDG methods, where the polynomial approximations are inherently discontinuous across cell boundaries. Furthermore, XDG methods are well suited for direct numerical simulations, since they are able to resolve all relevant flow phenomena, except the layer itself, in great detail with at the same time reasonable computational costs.

Shock-fitting　The idea of shock-fitting dates back to the 1940's, where Emmons (1944, 1948) calculated the transonic flow in a hyperbolic channel by solving the Euler equations by means of a velocity potential and a stream function in combination with a fitted shock wave. We recommend the textbook by Salas (2010) for a historical overview and a detailed introduction to shock-fitting.

Many early shock-fitting approaches are based on the discretization of the Euler equations by a Finite Difference Method (FDM) on a fixed rectangular reference domain, which is mapped onto a *time-dependent* physical domain. Starting from an initial guess of the shock position, a time marching procedure is applied in order to fit one boundary of the computational domain to the exact shape and position of the shock. The first practical solution to the supersonic blunt-body problem was published by Moretti and Abbett (1966), sparking several subsequent works (Moretti, 1969; Marconi and Salas, 1973; Pao and Salas, 1981; Moretti, 1987). We briefly introduce the original work by Moretti and Abbett (1966) in Section 6.2.

Variants of the original approach (Moretti and Abbett, 1966) were published by Zang et al. (1983), Hussaini et al. (1985), and Kopriva et al. (1991), who used Fourier and Chebyshev approximations. These approaches were also combined with multi-domain methods, where the entire domain of interest is decomposed into several sub-domains according to an a-priori knowledge about the flow (Wu and Zhu, 1996; Kopriva, 1991). The shocks are then fitted separately in each sub-domain. Recently, Romick and Aslam (2017) applied this methodology to detonation problems (Henrick, 2008; Short et al., 2008) by fitting a computational boundary to shock fronts and material interfaces. The authors demonstrated high-order convergence rates in the context of the two-dimensional reactive Euler equations.

---

[16]Extended version of Geisenhofer et al. (2020, Section 1).

Floating shock-fitting The main difficulty of shock-fitting approaches is the requirement of a sufficient a-priori knowledge about the flow. This motivated the development of so-called *floating shock-fitting* approaches, which additionally enable the treatment of internal shock boundaries (Moretti, 1976). In the context of FDMs, the movement of intersection points of the shock with the background grid was tracked (Moretti and Valorani, 1988; Di Giacinto and Valorani, 1989; Hartwich, 1991; Nasuti and Onofri, 1996). Rawat and Zhong (2010) obtained a high-order solution by using a fourth-order reconstruction of the shock front.

Another class of methods is based on a local adaption of the computational grid to fit the shock front (Morton and Paisley, 1989; Van Rosendale, 1994; Trépanier et al., 1996). Although they are conceptually different from the original floating shock-fitting approach, they are often added to this class of methods. Paciorri and Bonfiglioli (2009) proposed an approach which tracks the points on the shock front and remeshes the computational grid locally by using a constrained Delaunay triangulation so that the edges of the grid coincide with the shock front. In the bulk, a shock-capturing strategy is used to treat discontinuities which have not (yet) been fitted. Subsequent works with increasing physical and topological complexity were published by Paciorri and Bonfiglioli (2011) and Bonfiglioli et al. (2016).

In summary, there exist many combinations of shock-fitting and shock-capturing approaches in the literature. Mostly the shocks which are known a-priori are fitted and the remaining ones are captured. Consequently, this prohibits a global high-order convergence. Nevertheless, numerical experiments have shown that *partial* shock-fitting has a positive effect on the solution accuracy (Salas, 2010).

Shock detection in the context of shock-fitting Shock-fitting approaches inherently require sufficiently accurate information about the position and orientation of the shock front, whereas many shock-capturing approaches only detect the grid cells containing the shock, since sub-cell accuracy is not required. Early shock-fitting works are based on FDMs so that a suitable choice is to mark grid cells as containing a shock when characteristics could intersect within a single time-step (Moretti and Valorani, 1988; Di Giacinto and Valorani, 1989; Hartwich, 1991; Nasuti and Onofri, 1996). However, classical shock detectors in shock-capturing strategies, such as the modal-decay indicator by Persson and Peraire (2006) or the jump indicator by Dolejší et al. (2003) and Krivodonova et al. (2004) are also applied in the context of shock-fitting. For example, Van Rosendale (1994) detects the formation of shocks based on the density gradient, while Salas (2010) proposes several variants based on the pressure gradient. Usually, an over-sensitive shock detection leads to a strong deterioration of the solution quality in shock-capturing approaches. By contrast, such superfluous shocks typically disappear after a few time-steps in shock-fitting approaches so that an over-sensitive shock detection is less of an issue here.

6.2 The Origin of Shock-Fitting

In the 1950's, the development of supersonic aircraft was mostly driven by the military sector. A well-known single-engine supersonic aircraft is the Lockheed F-104 *Starfighter*, which was the first fighter designed for a sustained flight at Mach 2 (Anderson, 1990, Chapter 12). Another example is the reusable glider X-20 *Dyna-Soar* developed by Boeing in collaboration with the

United States Air Force and the National Aeronautics and Space Administration (NASA), which can be seen as one of the predecessors of the space shuttle program, for which the development was intensified in the end of the 1960's. During the 1950's, the blunt-body concept was introduced for hypersonic vehicles, minimizing aerodynamic heating, for example, by blunt noses and blunt leading edges. In this context, the supersonic blunt-body problem became an active subject of research for more than 15 years. At that time, no theoretical solutions for such a flow field existed. The most challenging part was the uniform treatment of the mixed subsonic-supersonic flow region behind the curved and detached shock wave in front of the blunt body by a single numerical technique. A major breakthrough was achieved by Moretti and Abbett (1966), who solved the *steady* blunt-body problem by means of an *unsteady* time-marching procedure, allowing the calculation of the mixed subsonic-supersonic flow field with a uniform technique. It is worthwhile mentioning that *unsteady* inviscid compressible flow is always *hyperbolic* in two and three spatial dimensions, and in any flow regime (Anderson, 1990, Chapter 11.3). This is the main advantage of the unsteady time-marching technique in contrast to other characteristic-based approaches, where, for example, in steady two-dimensional irrotational flow, the flow regimes are *elliptic* for M < 1, *parabolic* for M = 1, and *hyperbolic* for M > 1 due to the made simplifications. Beyond the original work, we recommend the summaries by Anderson et al. (1968), Anderson (1990, Section 12.5) and Salas (2010, Chapter 5) as an introduction to the blunt-body problem.

In the following, we briefly sketch the main principles of the time-marching procedure in Section 6.2.1, before introducing the concept of shock-fitting based on the fundamental work by Moretti and Abbett (1966) in Section 6.2.2.

6.2.1 Time-Marching Procedure

In the context of a steady flow, such as in the blunt-body problem, the time-marching procedure can be applied to drive the solution into the steady state by considering the unsteady equations.

In this section, we introduce the time-marching procedure based on Anderson (1990, Section 12.1) by exemplarily considering the quasi-one-dimensional flow of a calorically perfect gas through a convergent-divergent nozzle, as shown in Figure 6.1. In experiments, the flow through the nozzle is usually initiated by opening high-pressure tanks so that the initial conditions are often called *reservoir conditions* in the literature. For simplicity, several grid points on the mid-line of the nozzle are considered in the context of an FDM. The calculation is initialized with arbitrary values for all quantities at these points, except for the first point at the inlet where the reservoir conditions $(\cdot)_{\mathrm{rsv}}$ are given. We look for the steady-state solution of a general scalar quantity $\psi(\mathbf{x}, t)$, for example, the density ρ or the velocity u_1, at each grid point by means of a Taylor series expansion

$$\psi(\mathbf{x}, t + \Delta t) = \psi(\mathbf{x}, t) + \left(\frac{\partial \psi(\mathbf{x}, t)}{\partial t} \right)\Bigg|_{t} \Delta t + \ldots , \tag{6.1}$$

where the time-step size Δt has to fulfill an explicit time-step restriction, see Section 4.1.3. The temporal derivative $(\partial \psi / \partial t)|_t$ of Equation (6.1) has to be approximated in a suitable way in order to advance the solution in time. Before doing so, we state the conservation laws for

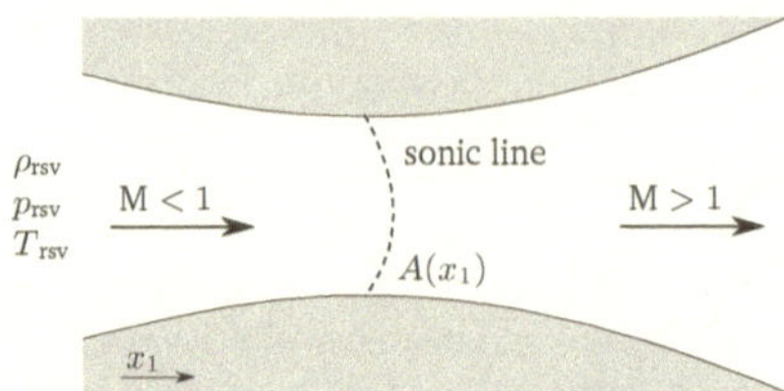

Figure 6.1: Quasi-one-dimensional flow through a convergent-divergent nozzle. A mixed subsonic-supersonic flow field establishes, where the flow is subsonic at the beginning (M < 1). The flow accelerates through the entire nozzle until it reaches the sonic limit (M = 1), where the cross-section $A(x_1)$ is minimal. Behind the sonic line, the flow is supersonic (M > 1). The reservoir conditions to drive the flow are denoted by $(\cdot)_{\mathrm{rsv}}$ (adapted from Anderson, 1990, Figure 12.4).

mass, momentum, and internal energy for unsteady quasi-one-dimensional flow in the form

$$\frac{\partial \rho}{\partial t} = -\frac{1}{A(x_1)} \frac{\partial(\rho u_1 A(x_1))}{\partial x_1}, \tag{6.2a}$$

$$\frac{\partial u_1}{\partial t} = -\frac{1}{\rho}\left(\frac{\partial p}{\partial x_1} + \rho u_1 \frac{\partial u_1}{\partial x_1}\right), \tag{6.2b}$$

$$\frac{\partial e}{\partial t} = -\frac{1}{\rho}\left(p\frac{\partial u_1}{\partial x_1} + \rho u_1 \frac{\partial e}{\partial x_1} + p u_1 \frac{\partial \ln(A(x_1))}{\partial x_1}\right), \tag{6.2c}$$

where $A(x_1)$ denotes the cross-section of the nozzle, which reduces to a line in the quasi-one-dimensional setting. The conservation laws (6.2) are supplemented with the perfect gas relations $p = \rho R_{\mathrm{gas}} T$ and $e = c_v T$, see Section 2.1.2. The temporal derivative $(\partial \psi / \partial t)|_t$ of the Taylor series expansion (6.1) can then be approximated by a first-order forward finite difference in time $(\partial \psi / \partial t)|_t \approx [\psi(t) + \psi(t + \Delta t)]/\Delta t$, which can be expressed by finite differences of the spatial derivatives on the right hand sides of the conservation laws (6.2). To achieve second-order accuracy, the problem is often solved by the MacCormack's scheme using a predictor-corrector procedure (MacCormack, 2003, orig. 1969).

In summary, the time-marching procedure delivers the steady-state solution for a mixed subsonic-supersonic unsteady quasi-one-dimensional flow by following the subsequent general steps (Anderson, 1991, Section 12.1):

(1) Arbitrary values for the physical quantities $\psi_0(\mathbf{x}) = \psi(\mathbf{x}, t_0)$ are set in the entire domain, except at the domain boundaries. In case of the convergent-divergent nozzle, the reservoir conditions $(\cdot)_{\mathrm{rsv}}$ are specified at the inlet.

(2) The quantities $\psi(\mathbf{x}, t)$ are advanced in time by means of the Taylor series expansion (6.1), which is terminated after the first-order term. The temporal derivative $(\partial \psi / \partial t)|_t$ is discretized, for example, by a forward finite difference or MacCormack's scheme (Mac-Cormack, 2003, orig. 1969). Furthermore, the supplemented conservation laws for unsteady quasi-one-dimensional flow (6.2) are employed.

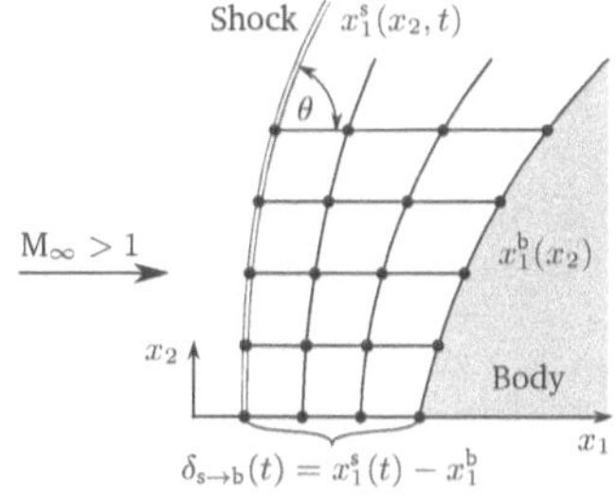

(a) Physical domain (x_1, x_2).

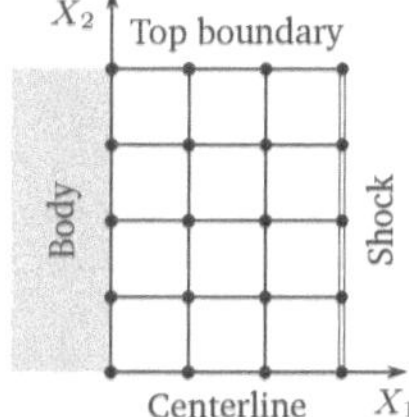

(b) Reference domain (X_1, X_2).

Figure 6.2: The supersonic blunt-body problem. The physical domain is mapped onto a rectangular reference domain, where the shock coincides with the right domain boundary. The supersonic inflow conditions ($M_\infty > 1$) and the body geometry $x_1^b(x_2)$ are given, whereas the initial shock position $x_1^s(x_2, t_0)$ has to be assumed (adapted from Moretti and Abbett, 1966, Figure 1; Anderson, 1990, Figures 12.10 and 12.11).

(3) Step (2) is repeated until the steady-state solution is reached so that $(\partial\psi((\mathbf{x}, t))/\partial t)|_t \ll 1$, for all $\mathbf{x} \in \Omega$.

6.2.2 The Supersonic Blunt-Body Problem

Moretti and Abbett (1966) published the first practical solution to the steady supersonic blunt-body problem by employing a time-marching procedure as presented in Section 6.2.1. The authors proposed an FDM with high accuracy for the uniform treatment of the mixed subsonic-supersonic flow field, where a bow shock stands in front of a blunt body.

In the subsequent years, the work of Moretti and Abbett (1966) gained much popularity due to the following properties: First, it is a *direct problem*, meaning that the body geometry is given. The counterpart is the *inverse problem*, where the shock shape is given and the body supporting this configuration is calculated, rendering it as a rather theoretical approach (Anderson, 1990, Section 12.5). Second, an arbitrarily high accuracy can be obtained, which only depends on the quality of the input data. Third, it was fast on a computer at that time, and sufficiently accurate results could be obtained even on a coarse grid. Furthermore, the shock wave is treated more realistically as a discontinuity and is not resolved over several grid cells (Bohachevsky and Rubin, 1966).

Figure 6.2a shows the physical computational domain (x_1, x_2), where a uniform supersonic inflow ($M_\infty > 1$) is prescribed and the body geometry is given. The shock front is described by $x_1^s = x_1^s(x_2, t)$ and the surface of the body is denoted by $x_1^b = x_1^b(x_2)$. The shock detachment distance is defined as $\delta_{s\to b}(t) = x_1^s(t) - x_1^b$. The angle between the shock tangent and the x_1-axis is denoted by θ. Since the problem is symmetric, only the upper half of the computational domain is considered. The physical domain (x_1, x_2) between the shock and the body is discretized by a numerical grid, which is then mapped onto a rectangular reference domain (X_1, X_2) by the transformation

$$X_1 := \frac{x_1 - x_1^{\mathrm{b}}}{\delta_{\mathrm{s} \to \mathrm{b}}}, \qquad X_2 := x_2, \tag{6.3}$$

as shown in Figure 6.2b. Moretti and Abbett (1966) use the transformation (6.3) onto a Cartesian uniform grid, since it considerably simplifies the calculations and additionally improves the solution accuracy in the context of an FDM (Ferziger and Perić, 2008, Chapter 3). The shock velocity is defined as $u_{\mathrm{s}}(t) = \mathrm{d}x_1^{\mathrm{s}}/\mathrm{d}t$, since the position and the shape of the shock $x_1^{\mathrm{s}}(x_2, t)$ will vary until the steady state is reached. Note that the shock is a boundary of the computational domain, which is common for classical shock-fitting approaches.

The governing equations of the blunt-body problem are the Euler equations for two-dimensional isentropic flow, see also Equation (2.5), which we state here for the sake of completeness in the form

$$\frac{\partial \rho}{\partial t} + \frac{\partial (\rho u_1)}{\partial x_1} + \frac{\partial (\rho u_2)}{\partial x_2} = 0, \tag{6.4a}$$

$$\rho \frac{\partial u_1}{\partial t} + \rho u_1 \frac{\partial u_1}{\partial x_1} + \rho u_2 \frac{\partial u_1}{\partial x_2} = -\frac{\partial p}{\partial x_1}, \tag{6.4b}$$

$$\rho \frac{\partial u_2}{\partial t} + \rho u_1 \frac{\partial u_2}{\partial x_1} + \rho u_2 \frac{\partial u_2}{\partial x_2} = -\frac{\partial p}{\partial x_2}, \tag{6.4c}$$

$$\frac{\mathrm{D}\hat{s}}{\mathrm{D}t} = \frac{\mathrm{D}(p/\rho^{\gamma})}{\mathrm{D}t} = 0, \tag{6.4d}$$

where we use the isentropic assumption for a calorically perfect gas, $p/\rho^{\gamma} = \mathrm{const.}$, in order to reformulate the energy equation, see also Equation (2.19). In the next steps, auxiliary variables are introduced and the governing equations (6.4) are non-dimensionalized with the free-stream conditions. The reader is referred to the original work (Moretti and Abbett, 1966) for a detailed derivation. As illustrated in Figure 6.2b, the reference domain (X_1, X_2) can be split into several sets of points: the body points $(X_1 = 0)$, the shock points $(X_1 = 1)$, the top boundary $(X_2 = X_{2,\max})$, and the interior points, which also include the centerline $(X_2 = 0)$.

The governing equations (6.4) are then solved by means of a time-marching procedure, see Section 6.2.1. For this, the initial shock shape $x_1^{\mathrm{s}}(x_2, t_0)$ has to be guessed. The values at the shock points $(X_1 = 1)$ are prescribed via the moving shock relations (2.37), whereas arbitrary values can be assumed at all other grid points. In every iteration $t \to t + \Delta t$, new values of the physical state variables, the shock detachment distance $\delta_{\mathrm{s} \to \mathrm{b}}(t)$, and the shock shape $x_1^{\mathrm{s}}(x_2, t)$ are computed.

An essential variable is the shock velocity $u_{\mathrm{s}}(\mathbf{x}, t)$, which is not known in the beginning. We add the dependency on the spatial vector $\mathbf{x}$, since not only the position of the shock but also its shape changes. Even though the underlying problem is steady, the shock will move until its correct position is reached, which corresponds to the steady-state solution. Consequently, an iteration procedure for $u_{\mathrm{s}}(\mathbf{x}, t + \Delta t)$ is required at each new time-level. For that, an independent solution is needed in order to judge the quality of the obtained value for $u_{\mathrm{s}}(\mathbf{x}, t + \Delta t)$. This is achieved by means of a characteristics technique, where a comparison solution is propagated from interior points towards the shock, as shown in Figure 6.3b. For this, at each shock point $(X_1 = 1)$ a local coordinate system $(\hat{\xi}, \hat{\eta})$, aligned normal and tangential to the shock, is introduced, see Figure 6.3a. Along the normal direction $\hat{\xi}$, the quasi-one-dimensional equations (6.2) can be

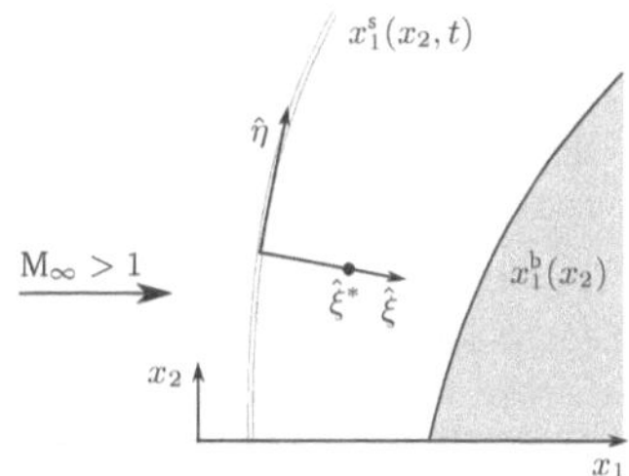

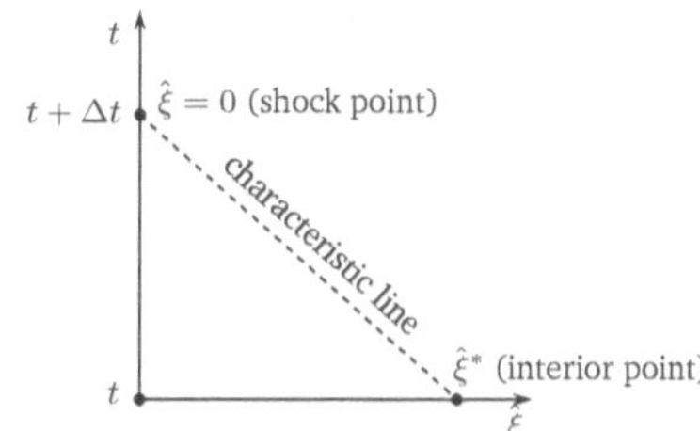

(a) Local coordinate system $(\hat{\xi}, \hat{\eta})$ at a shock point $(\hat{\xi} = 0)$ with an interior point $\hat{\xi}^*$.

(b) A one-dimensional characteristic which emanates from an interior point $\hat{\xi}^*$ determines the quantities at the shock point $(\hat{\xi} = 0)$.

Figure 6.3: The supersonic blunt-body problem. The physical quantities at a shock point $(\hat{\xi} = 0)$ are obtained by projecting the solution of the modified quasi-one-dimensional governing equations (6.2) along a characteristic line from an interior point $\hat{\xi}^*$. In particular, the correct shock velocity $u_s(\hat{\xi} = 0, t + \Delta t)$ can be obtained via an iterative procedure until it matches the moving shock relations (2.37) (adapted from Moretti and Abbett, 1966, Figures 3 and 4; Anderson, 1990, Figures 12.12 and 12.13).

applied supplemented with forcing terms containing derivatives in tangential direction. From an interior point $\hat{\xi}^*$ in the flow field where the solution is already known, the solution at the shock point $(\hat{\xi} = 0)$ can be obtained, since both points lie on the same characteristic line, see Figure 6.3b. For that, Moretti and Abbett (1966) additionally consider the compatibility equation for the characteristic $\mathrm{d}\hat{\xi}/\mathrm{d}t = u_{\hat{\xi}} - a$, where $u_{\hat{\xi}}$ denotes the component of the velocity vector in $u_{\hat{\xi}}$-direction. This procedure is repeated until the correct shock speed $u_s(\hat{\xi} = 0, t + \Delta t)$, and, thus, the physical quantities at the shock point are obtained.

For the body points $(X_1 = 0)$ the calculation is simpler, since the entropy on the body is known. The entropy on the entire body $\hat{s}(x_1^b(x_2), x_2)$ is defined by the entropy at the stagnation point $\hat{s}(x_1^b(0), 0)$. This value is obtained by evaluating the normal shock relations (2.30) along the stagnation line $(x_1, 0)$. We skip further details here and refer to the original work (Moretti and Abbett, 1966). All presented steps are repeated until the time-marching procedure converges to the steady-state solution.

Note on the application in discontinuous Galerkin methods The results of a standard DG method in combination with an artificial viscosity based shock-capturing strategy, see Chapter 5, show the following: High-order polynomial solutions immediately start to oscillate in the vicinity of discontinuities. As a remedy, discontinuities may be smoothed over a length scale of $O(h/P)$ by applying artificial viscosity, which leads to the loss of high-order convergence. In order to regain it, the idea of fitting one boundary of the computational domain or an additional interface to the shock front seems promising in the context of DG methods at first glance, since they inherently solve local Riemann problems across cell boundaries. The crucial point of the original approach by Moretti and Abbett (1966) is the application of an unsteady time-marching procedure to solve the steady-state problem. However, an adaption to DG methods is challenging for the following reasons. The iterative correction procedure for the position and shape of the shock prevents the one-to-one applicability. As soon as the current shock position does not coincide with the steady-state solution, the polynomial solution starts to oscillate and pollutes the solution downstream of the shock. Therefore, the iterative correction procedure which is based on a physically reasonable, converged solution at the interior points will fail.

Shock-fitting in DG methods is applicable if the exact shock shape is already known such as in simple configurations containing normal shock waves. For general two-dimensional applications, the shape and the position of the shock are complex and not known a-priori. Thus, suitable reconstruction techniques of the shock front are essentially needed for the general application in a high-order XDG method so that the shock front can be located in the computational domain without limitation. In this work, we discuss a novel reconstruction procedure of the shock front in Sections 6.3 and 6.4.

6.3 Sharp Reconstruction of the Shock Front

Discontinuities are always smeared over a width of $O(h/P)$ when using shock-capturing strategies based on artificial viscosity, see Chapter 5. Furthermore, the application of artificial viscosity spoils the high-order convergence rate of DG methods. In this and the subsequent section, we aim for solving this issue by means of a sharp reconstruction of the shock front, which is essential in the context of a high-order XDG method for supersonic compressible flow. The shock front and the surface of solid bodies are represented by the zero iso-contours of two different level-set functions.

In Section 6.3.1, we verify the *XDGShock* solver of the *BoSSS* framework for the application of two level-set functions in combination with the cell-agglomeration technique presented in Section 3.3.2. In Sections 6.3.2 and 6.3.3, we introduce auxiliary methods for the sharp reconstruction of the shock front, which we derive in Section 6.3.4.

In order to motivate the methodology presented in this section, we show a discontinuous Galerkin immersed boundary method (DG IBM) simulation of a supersonic blunt-body problem with a polynomial degree of $P = 2$ in Figure 6.4. We apply the two-step shock-capturing strategy, which we have investigated in detail in the previous chapter. The blunt body is represented by a rounded rectangle which is composed of the quadrants of two circles with the center coordinates $(x_{1,c_1}, x_{2,c_1})^\mathsf{T} = (0.0, 0.5)^\mathsf{T}$ and $(x_{1,c_2}, x_{2,c_2})^\mathsf{T} = (0.0, -0.5)^\mathsf{T}$, respectively, and the radii $r_{c_1} = r_{c_2} = 0.5$ as well as a vertical line segment at $x_1 = -0.5$. For the representation of the blunt body, we define a continuous level-set function $\varphi_1(\mathbf{x}) \in C^2(\Omega)$ in its quadratic formulation

$$\varphi_1(\mathbf{x}) = \begin{cases} (x_1 - x_{1,c_1})^2 + (x_2 - x_{2,c_1})^2 - r_{c_1}^2\,, & \text{if } x_2 \geq 0.5\,, \\ (x_1 - x_{1,c_2})^2 + (x_2 - x_{2,c_2})^2 - r_{c_2}^2\,, & \text{if } x_2 < -0.5\,, \\ x_1^2 - (-0.5)^2\,, & \text{otherwise} \end{cases} \tag{6.5}$$

so that the zero iso-contour $\mathfrak{I}_1 = \{\mathbf{x} \in \Omega : \varphi_1(\mathbf{x}) = 0\}$ exactly represents the blunt body in the DG approximation space $\mathbb{P}_{P=2}^\mathsf{X}(\mathcal{K}_h)$. We prescribe supersonic free-steam conditions with a Mach number of $\mathrm{M}_\infty = 4.0$ at the left, top, and bottom boundary as well as supersonic outlet

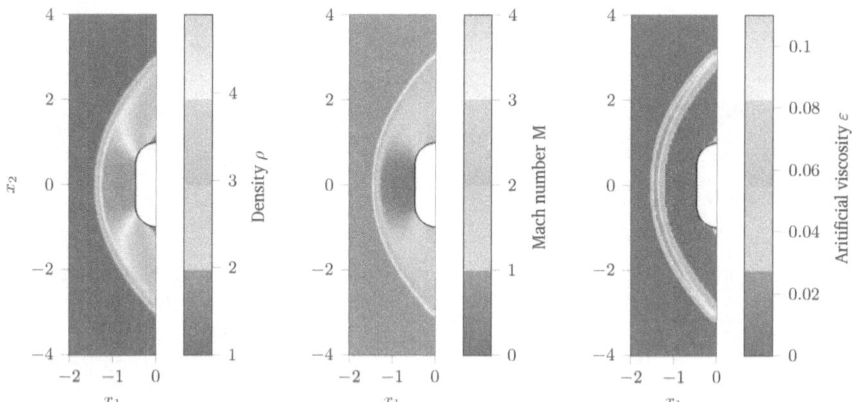

Figure 6.4: Bow shock computed with a DG IBM with shock-capturing. The shock is smeared over several cells due to the application of artificial viscosity. The blunt body is represented by the zero iso-contour of a level-set function. Free-stream conditions with a Mach number of $M_\infty = 4.0$ are prescribed.

conditions at the right boundary. Note that supersonic-inlet boundary conditions are applicable at the top and bottom boundary, since both are located entirely in the supersonic flow regime where all characteristics move downstream. We apply adiabatic slip-wall boundary conditions at the blunt body. This test case is based on the setting proposed by the HiOCFD5-Workshop (2017). The computational domain is set to $\Omega = [-2 - \varepsilon_x, 0 - \varepsilon_x] \times [-4, 4]$, with a grid consisting of 40×160 cells with $h = 0.05$. We slightly shifted the domain with $\varepsilon_x = h/2$ in the negative x_1-direction in order to produce a reasonable cut configuration where the interface does not coincide with the straight vertical part of the blunt body at $x_1 = -0.5$. A different, 'more suitable' choice of the grid could considerably reduce the number of cut-cells and improve their shape, respectively. The shock-capturing parameters are set to $S_0 = 1.0 \cdot 10^{-3}$, $\varepsilon_0 = 1.0$, $\kappa = 1.0$, and $\lambda_{c,\max} = 15$. The cell-agglomeration threshold is given by $\delta_{\text{agg}} = 0.3$. The explicit Euler scheme is used for time-integration, since we do not focus on temporal efficiency in this test case. In Figure 6.4, we show the converged density, Mach number, and artificial viscosity distributions at $t_{\text{end}} = 16.0$. The plots clearly indicate that the strong shock ($M_\infty = 4.0$) is smeared out by the two-step shock-capturing strategy, see Section 5.4.

6.3.1 Verification of Two Level-Set Functions

In this section, we verify the application of two level-set functions in the *XDGShock* solver of the *BoSSS* framework. The zero iso-contours of the level-set function represent the solid body and the shock front, respectively. Furthermore, we use the cell-agglomeration technique for improving the conditioning of the system matrix and for loosening the time-step restrictions in the context of explicit time-integration schemes, see Sections 3.3.2 and 4.1, respectively.

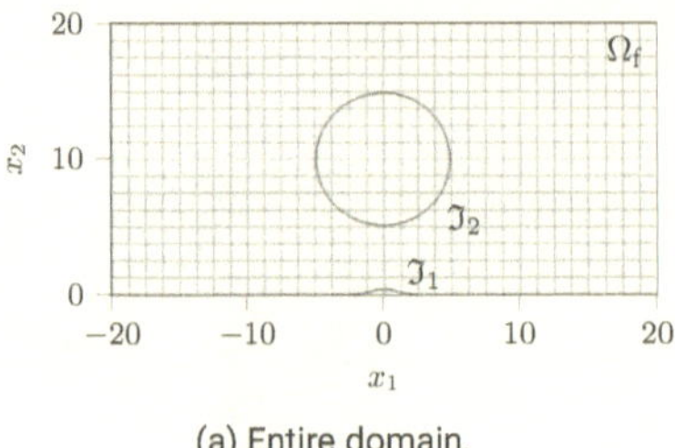
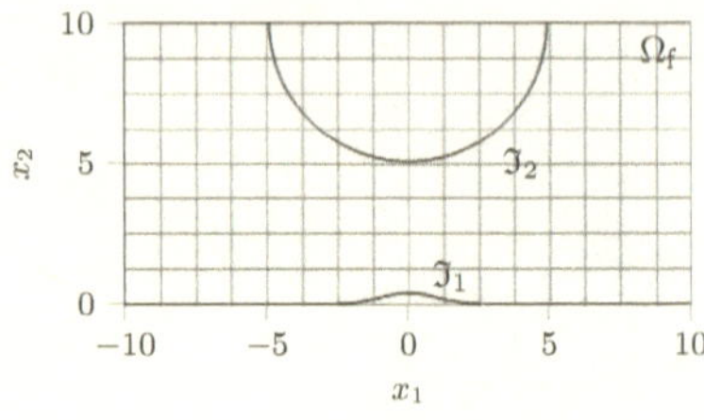

(a) Entire domain.   (b) Zoom-in around the bump.

Figure 6.5: Computational domain of a Gaussian bump in the context of an XDG method. The interface $\Im_1$ describes the smooth bump by means of adiabatic slip-wall boundary conditions. The interface $\Im_2$ represents an 'artificial' circle where standard convective fluxes are evaluated in order to verify the implementation of two level-set functions. The fluid domain is denoted by $\Omega_f = \{\mathbf{x} \in \Omega : \varphi_1(\mathbf{x}) > 0\}$.

Subsonic flow over a Gaussian bump We consider the subsonic flow over a smooth Gaussian bump. This test case was also used for the verification of the modified Hierarchical Moment-Fitting (HMF) quadrature rules for subsonic compressible flow (Müller et al., 2013; Müller et al., 2017) in combination with the cell-agglomeration technique. They are implemented in the *Compressible Navier-Stokes (CNS)* solver of the *BoSSS* framework. The Gaussian bump is represented by the zero iso-contour $\Im_1 = \{\mathbf{x} \in \Omega : \varphi_1(\mathbf{x}) = 0\}$ of a first level-set function

$$\varphi_1(\mathbf{x}) = x_2 - 0.01 + \frac{1}{\sqrt{2\pi}} e^{-0.5x_1^2} . \tag{6.6}$$

Additionally, an 'artificial' circle, which does not represent a solid body and is only used for verifying the implementation of two level-set functions, is given by the zero iso-contour $\Im_2 = \{\mathbf{x} \in \Omega : \varphi_2(\mathbf{x}) = 0\}$ of a second level-set function $\varphi_2(\mathbf{x})$ in its quadratic formulation

$$\varphi_2(\mathbf{x}) = - \left[(x_1 - 0.0)^2 + (x_2 - 10.0)^2 - 4.9^2\right] . \tag{6.7}$$

Both level-set functions $\varphi_1(\mathbf{x})$ and $\varphi_2(\mathbf{x})$ are not time-dependent. They are discretized by a high-order DG approximation $P_\varphi = 8$ in order to not introduce errors, in particular by the representation of the Gaussian bump. We choose $P_\varphi = 8$ for both level-set functions for implementation reasons. We prescribe adiabatic slip-wall boundary conditions at the Gaussian bump $\Im_1$, whereas standard convective fluxes are evaluated at the circle $\Im_2$. Figure 6.5 shows an exemplary coarse grid with 32×16 cells on the computational domain $\Omega = [-20, 20] \times [0, 20]$ with $h = 1.25$. We initialize the flow field with free-stream conditions with a Mach number of $M_\infty = 0.5$ and a heat capacity ratio of $\gamma = 1.4$ by

$$(\rho_\infty, u_{1,\infty}, u_{2,\infty}, p_\infty)^\mathsf{T} = \left(1, 1, 0, \underbrace{\frac{\rho u_{\infty,1}^2}{M_\infty^2 \gamma}}_{\approx 2.857}\right)^\mathsf{T} \tag{6.8}$$

and additionally prescribe these values at all domain boundaries.

We compare our numerical solutions with the free-stream entropy $\hat{s}_\infty = p_\infty/\rho_\infty^\gamma$, since no analytical solution of the flow field exists. In this test case, $\hat{s}_\infty \approx 2.857$ holds in the entire

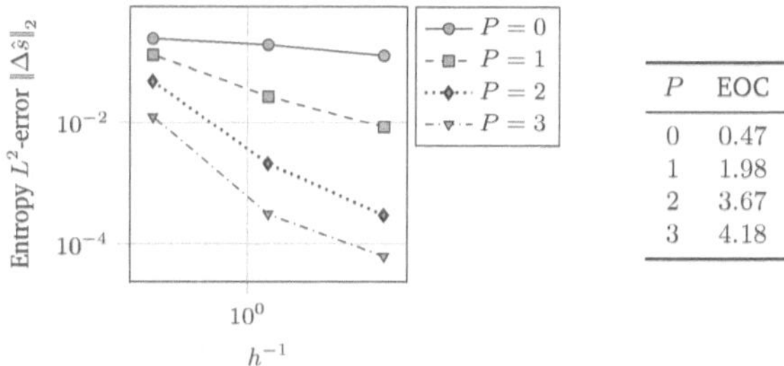

Figure 6.6: Subsonic flow over a Gaussian bump computed with an XDG method using two level-set functions. Spatial convergence study for free-stream conditions with a Mach number of $M_\infty = 0.5$. The entropy L^2-error $\|\Delta \hat{s}\|_2 = \|p/\rho^\gamma - \hat{s}_\infty\|_2$ is calculated with respect to the free-stream entropy $\hat{s}_\infty$, see Equation (6.9). EOC denotes the experimental order of convergence.

flow field. Consequently, a deviation from this state can be used as a measure for the accuracy of the numerical method. We introduce the entropy L^2-error $\|\Delta \hat{s}\|_2$ in the fluid domain $\Omega_f = \{\mathbf{x} \in \Omega : \varphi_1(\mathbf{x}) > 0\}$ as

$$\|\Delta \hat{s}\|_2 = \sqrt{\int_{\mathbf{x} \in \Omega_f} \Delta \hat{s}^2 \, dV} \stackrel{\text{Def. 3.1}}{=} \sqrt{\int_{\mathbf{x} \in \Omega_f} \left(\frac{p}{\rho^\gamma} - \hat{s}_\infty \right)^2 dV}. \tag{6.9}$$

We perform a spatial convergence study for several polynomial degrees $P = \{0, 1, 2, 3\}$ and discretize the computational domain Ω into a set of equidistant Cartesian grids consisting of 32×16, 64×32, and 128×64 cells. We enable the cell-agglomeration technique at both interfaces $\mathfrak{I}_1$ and $\mathfrak{I}_2$ by using an agglomeration threshold of $\delta_{\text{agg}} = 0.3$. Figure 6.6 shows the results of the spatial convergence study, confirming the characteristic high-order convergence of DG methods for smooth solutions. The reader is referred to Figure 14 of the work by Müller et al. (2017) to approve the presented results with their spatial convergence study where only one level-set function for the representation of the Gaussian bump is used.

6.3.2 Patch-Recovery Filters

During the reconstruction procedure of the shock front, we apply patch-recovery filters as proposed in the work by Kummer and Warburton (2016). We briefly introduce their concept in the following paragraph.

Kummer and Warburton (2016) extended the work by Zienkiewicz and Zhu (1995), who obtained super-convergent stress and gradient values by a post-processing routine on sample points with super-convergent properties. In particular, Kummer and Warburton (2016) employed patch-recovery filters to stabilize the computation of the curvature at interfaces between two fluids in context of an XDG method. In their approach, the interface was sharply represented by a level-set formulation as in this work. In contrast to Zienkiewicz and Zhu (1995), who used a nodal projection for the patch-recovery operator, Kummer and Warburton (2016) applied a modal L^2-projection, allowing patches of arbitrary shapes. This is beneficial if the patch-recovery algorithm should only be executed on cut-cells or cells inside the narrow band of the interface, which contains all cut-cells and its neighbors. We repeat several definitions for convenience:

- A cut-cell $K_{j,\mathfrak{s}} \in \mathcal{K}_h^{\mathsf{X}}$ is defined as $K_{j,\mathfrak{s}} := K_j \cap \mathfrak{s}$ for a species $\mathfrak{s} \in \{\mathfrak{A}, \mathfrak{B}\}$, see Equation (3.67).

- A background cell $K_j \in \mathcal{K}_h$ is denoted as cut by the interface if $\oint_{K_j \cap \mathfrak{I}} 1 \, \mathrm{d}S \neq 0$. Furthermore, it is the union of both cut-cells ($K_j = K_{j,\mathfrak{A}} \cup K_{j,\mathfrak{B}}$).

- A cell $K_k \in \mathcal{K}_h^{\mathsf{X}}$ is a neighbor of a cell $K_j \in \mathcal{K}_h^{\mathsf{X}}$ if they share at least one point ($\overline{K}_k \cap \overline{K}_j \neq \emptyset$). This definition is different from the usual definition where neighboring cells share a common edge so that a quadrilateral cell on a Cartesian grid has eight neighbors $K_{k=1,\dots,8}$ in this case.

- We denote the set of all cut-cells, which is the same as the set of all background cells cut by the interface,

$$\mathcal{K}_h^{cc,0} := \left\{ K_j \in \mathcal{K}_h; \text{ if } \oint_{K_j \cap \mathfrak{I}} 1 \, \mathrm{d}S \neq 0 \right\}. \tag{6.10}$$

- We define the set of all cut-cells and their neighbors as

$$\mathcal{K}_h^{cc,1} := \left\{ K_j \in \mathcal{K}_h; \text{ if } \oint_{K_j \cap \mathfrak{I}} 1 \, \mathrm{d}S \neq 0 \text{ or } \oint_{K_k \cap \mathfrak{I}} 1 \, \mathrm{d}S \neq 0 \right\}. \tag{6.11}$$

The set of cells (6.11) is called the *narrow band*.

An exemplary cut configuration is shown in Figure 6.7, which also serves as the starting point for further definitions. We define a general patch-recovery operator prc_w on the set of cells $\mathcal{K}_h^{cc,w}$ as

$$\mathrm{prc}_w : \mathbb{P}_{\tilde{q}}(\mathcal{K}_h^{cc,w}) \to \mathbb{P}_{\tilde{r}}(\mathcal{K}_h^{cc,w}), \tag{6.12}$$

where $w \in \mathbb{N}^0$ denotes the width of the patch, and $\tilde{q}$ and $\tilde{r}$ with $\tilde{q} \leq \tilde{r}$ denote the polynomial degrees of the domain and co-domain, respectively. The operator (6.12) is defined as the L^2-projection onto a broken polynomial space $\mathbb{P}_{\tilde{r}}(\{\tilde{Q}_{K_j}\})$ of a so-called composite cell $\tilde{Q}_{K_j} := K_j \cup K_{k=1} \cup \dots \cup K_{k=K}$, consisting of a cell $K_j \in \mathcal{K}_h^{cc,w}$ and all of its neighbors K_k. For a cell $K_l \in \mathcal{K}_h^{cc,w}$, we define the patch-recovery operation $\mathrm{prc}_w(\psi) =: \vartheta$ as the L^2-projection of a polynomial field $\psi \in \mathbb{P}_{\tilde{q}}(\mathcal{K}_h^{cc,w})$ onto the space $\mathbb{P}_{\tilde{r}}(\{\tilde{Q}_{K_j}\})$ by

$$\vartheta|_{K_l} := \begin{cases} \mathrm{Proj}_{\mathbb{P}_{\tilde{r}}(\{\tilde{Q}_{K_j}\})}(\psi|_{\tilde{Q}_{K_j}})\Big|_{K_l}, & \text{if } K_l \in \mathcal{K}_h^{cc,w}, \\ 0, & \text{if } K_l \neq \mathcal{K}_h^{cc,w}. \end{cases} \tag{6.13}$$

In other words, for each cell $K_l \in \mathcal{K}_h^{cc,w}$, a projection of a broken polynomial is performed onto polynomials which are continuous in the composite cell $\tilde{Q}_{K_j}$. After applying the projection operation in Equation (6.13), the result $\vartheta|_{K_l}$ is obtained by restricting the polynomial in the composite cell $\tilde{Q}_{K_j}$ to the considered cell K_l. For the definition of multiple sweeps of the patch-recovery operation and further details, we refer to the original work by Kummer and Warburton (2016).

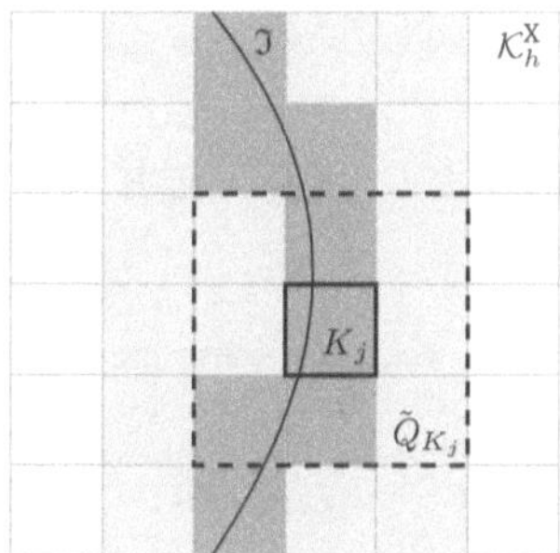

Figure 6.7: Patch-recovery filters. The set of all cut-cells is denoted by $\mathcal{K}_h^{cc,0} \subset \mathcal{K}_h^{\mathrm{X}}$ (dark gray) and the set additionally incorporating all neighboring cells by $\mathcal{K}_h^{cc,1} \subset \mathcal{K}_h^{\mathrm{X}}$, which is called the *narrow band* (dark and light gray). The composite cell $\tilde{Q}_{K_j}$ is defined as the set of cells which consists of a background cell K_j cut by the interface $\mathfrak{I}$ and all of its neighbors $K_{k=1,\dots,8}$ which are connected by at least one common point. The patch-recovery operation is performed by a projection onto the polynomial space $\mathbb{P}_{\tilde{r}}(\{\tilde{Q}_{K_j}\})$, see also Equation (6.13).

6.3.3 Continuity Projection

The approximation of a level-set function and, thus, of the interface, should be continuous in order to obtain a stable numerical method. In general, this condition is not satisfied by the projection onto a broken polynomial DG space. As a solution, we present an L^2-projection with constraints in order to enforce continuity across cell boundaries, following the derivation by Smuda (2020) and Smuda and Kummer (2020). For this, it is sufficient to ensure a continuous solution solely at several points on the inner edges Γ_{int} of a grid $\mathcal{K}_h$. As an example, we look at a polynomial basis function in a two-dimensional polynomial space, $\phi \in \mathbb{P}_P^{D=2}(\mathcal{K}_h)$. In this case, an inner edge may be represented by a one-dimensional basis function $\phi^* \in \mathbb{P}_P^{D=1}(\mathcal{K}_h)$. Using this, we can obtain a continuous polynomial solution on the entire grid $\mathcal{K}_h$ by modifying the numerical approximation at a sufficient number of points at the inner edges by means of additional equality constraints. In general, we look for the optimal L^2-projection of a broken polynomial field $\psi \in \mathbb{P}_P(\mathcal{K}_h)$ onto a continuous space $\mathbb{P}_{P*}(\mathcal{K}_h) \cap C^0(\Omega)$ with a degree of $P^* \geq P$. The corresponding projection operator is defined by

$$\Pi^{\mathbb{P}_{P*}} : \mathbb{P}_P(\mathcal{K}_h) \to \mathbb{P}_{P*}(\mathcal{K}_h),$$
$$\psi \mapsto \psi^{\mathbb{P}_{P*}}, \tag{6.14}$$

with the essential property $\langle \psi - \psi^{\mathbb{P}_{P*}}, \vartheta \rangle = 0, \forall \vartheta \in \mathbb{P}_{P*}$. The continuity projection is defined as the solution to a quadratic optimization problem with additional equality constraints for all inner edges $\Gamma_i \subseteq \Gamma_{\mathrm{int}}$, which can be written as

$$\min \left\| \psi^{C^0} - \psi^{\mathbb{P}_{P*}} \right\|_2^2, \qquad \text{on } \mathcal{K}_h, \tag{6.15a}$$

$$\text{such that} \quad \psi_{j,\mathrm{in}}^{C^0}\Big|_{\Gamma_i} = \psi_{j,\mathrm{out}}^{C^0}\Big|_{\Gamma_i}, \qquad \Gamma_i \subseteq \Gamma_{\mathrm{int}}. \tag{6.15b}$$

Here, the continuous solution is denoted by $\psi^{C^0} = \mathbf{x}_{cp}^{\mathsf{T}}\boldsymbol{\phi}$, where $\mathbf{x}_{cp}$ is the unknown coordinate vector and $\psi^{\mathbb{P}^{P^*}} = \mathbf{b}_{cp}^{\mathsf{T}}\boldsymbol{\phi}$ is the result of the projection (6.14). The optimization problem (6.15) is then reformulated as

$$\hat{J}(\mathbf{x}_{cp}) = \frac{1}{2}\mathbf{x}_{cp}^{\mathsf{T}}\mathbf{M}\mathbf{x}_{cp} - \mathbf{x}_{cp}^{\mathsf{T}}\mathbf{b}_{cp} + \mathbf{b}_{cp}^{\mathsf{T}}\mathbf{b}_{cp} \to \min, \tag{6.16a}$$

$$\text{such that} \quad \mathbf{A}_c\mathbf{x}_{cp} = 0, \tag{6.16b}$$

where $\hat{J}(\mathbf{x}_{cp})$ is the objective function. The mass matrix is equal to the identity matrix ($\mathbf{M} = \mathbf{I} \in \mathbb{R}^{N(P^*),N(P^*)}$), since $\boldsymbol{\phi}$ is an orthonormal basis. The matrix $\mathbf{A}_c \in \mathbb{R}^{N_c,N(P^*)}$ denotes the constraint matrix, and N_c denotes the number of constraints at the inner edges Γ_{int}. For the incorporation of the equality constraints and further details, the reader is referred to the original work by Smuda (2020).

6.3.4 Shock Level-Set Reconstruction

In this section, we derive an algorithm for the reconstruction of a sharp shock front, or sharp shock interface, respectively, by means of a level-set function. For that, we reuse the stationary bow shock in front of a blunt body as a show case, see the introduction of this chapter. In particular, we assume that the inflection points of the smoothed density field, which is obtained by a DG IBM computation with shock-capturing, are a sufficiently good approximation of the shock front. First, we compute a cell-accurate approximation of the shock position by a point-based search of the inflections points $\mathcal{X}_{\text{infl}}(\rho) = \{\mathbf{x}_{\text{infl},1},\ldots,\mathbf{x}_{\text{infl},M}\}$ of the density field ρ. Note that the shape and the position of the shock front is not known analytically for a general, non-trivial higher-dimensional application. By contrast, the presented reconstruction procedure can be used in the general case. Second, we reconstruct a level-set function based on $\mathcal{X}_{\text{infl}}(\rho)$. Finally, the reconstructed level-set function is filtered and optimized with the methodology presented in Sections 6.3.2 and 6.3.3.

Computing candidate points In general, artificial viscosity ε is added to cells $K_j \in \mathcal{K}_h^{\mathsf{X}} \subset \Omega_{\mathsf{f}}$, where the smoothness of the polynomial approximation is below a certain threshold, see Section 5.4. Thus, cells with $\varepsilon_j > 0$ contain parts of the shock front with a high probability so that we seed points $\mathcal{X}_{\text{seed}}(\varepsilon) = \{\mathbf{x}_{\text{seed},1},\ldots,\mathbf{x}_{\text{seed},K}\}$ at their centers. Based on the seeding points $\mathcal{X}_{\text{seed}}(\varepsilon)$, a point-based search of possible candidate points $\mathcal{X}_{\text{cand}}(\rho) = \{\mathbf{x}_{\text{cand},1},\ldots,\mathbf{x}_{\text{cand},L}\}$ of the inflection points $\mathcal{X}_{\text{infl}}(\rho)$ is initialized according to Algorithm 6.1. Note that the auxiliary method FindCandidatePoint($\ldots$) only returns candidate points which have converged under a user-specified number of iterations $N_{\max}$. In Figure 6.8, we show the seeding points $\mathcal{X}_{\text{seed}}(\varepsilon)$ and the candidate points $\mathcal{X}_{\text{cand}}(\rho)$. Non-converged points have already been removed. The corresponding density and artificial viscosity distributions have been shown in Figure 6.4 in the beginning of this chapter.

Sorting and clustering the candidate points Figure 6.9 shows the results of the auxiliary method SortAndCluster($\mathcal{X}_{\text{cand}}$) in Algorithm 6.1. First, we cluster the candidate points $\mathcal{X}_{\text{cand}}(\rho)$ by the K-means algorithm (Macqueen, 1967) according to their density values, since a shock is governed by a sudden change of thermodynamic properties such as the density. In this test case, we obtain a clustering consisting of three cell clusters $\mathcal{M}^\rho = \{M_1^\rho, M_2^\rho, M_3^\rho\}$,

Algorithm 6.1: Pseudocode for determining the inflection points $\mathcal{X}_{\mathrm{infl}}(\rho)$ of the density field ρ. A set of points $\mathcal{X}_{\mathrm{infl}}(\rho) = \textsc{FindShockPosition}(\rho, \varepsilon, N_{\mathrm{max}})$ is computed by a point-based search. We seed points $\mathbf{x}_{\mathrm{seed}} \in \mathcal{X}_{\mathrm{seed}}(\varepsilon)$ at the center of cells $K_j \in \mathcal{K}_h^{\mathrm{X}} \subset \Omega_{\mathrm{f}}$ where artificial viscosity is active ($\varepsilon_j > 0$).

```
 1: function FindShockPosition(Density ρ, Artificial viscosity ε, Iterations N_max)
 2:     X_seed ← {}                                          ▷ Initiate seeding points
 3:     for all K_j ∈ K_h^X ⊂ Ω_f do                         ▷ Loop over cells in fluid domain
 4:         if ε_j > 0 then                                  ▷ Find cells with artificial viscosity
 5:             X_seed ← X_seed ∪ {x_j,center}               ▷ Add coordinates of cell center
 6:         end if
 7:     end for
 8:
 9:     X_cand ← {}                                          ▷ Initiate candidate points
10:     for all x_seed ∈ X_seed do                           ▷ Loop over seeding points
11:         X_cand ← X_cand ∪ FindCandidatePoint(x_seed, ρ, N_max)   ▷ Add cand. point if converged
12:     end for
13:
14:     X_infl ← SortAndCluster(X_cand)   ▷ See paragraph Sorting and clustering the candidate points
15:
16:     return X_infl                                        ▷ Return inflection points
17: end function
18:
19:
20: function FindCandidatePoint(Seeding point x_seed, Density ρ, Iterations N_max = 100)
21:     PatchRecovery(Gradient(ρ))                           ▷ See Section 6.3.2
22:     PatchRecovery(Hessian(ρ))                            ▷ Optimize derived fields once for later evaluation
23:     n ← 1                                                ▷ Initiate iteration count
24:     x_cand ← x_seed                                      ▷ Initiate return value
25:     s = 0.5 h                                            ▷ Initiate step size with 0.5 of characteristic grid size
26:
27:     while n ≤ N_max do                                   ▷ Iterate until max. iter. are reached or solution is found
28:         x_cand ← x_cand + ∇ρ(x_cand) s                   ▷ Search along the density gradient
29:
30:         if sign(ρ''(x_cand))_n ≠ sign(ρ''(x_cand))_{n-1} then   ▷ Check sign of sec. derivative along grad.
31:             s ← -s/2                                     ▷ Bisection of the step size and change of direction
32:         end if
33:
34:         if |(x_cand)_n - (x_cand)_{n-1}| < 10^{-12} then ▷ Termination criterion
35:             return (x_cand, converged = true)            ▷ Return converged solution
36:         end if
37:         n ← n + 1
38:     end while
39:
40:     return (x_cand, converged = false)                   ▷ Return non-converged solution
41: end function
```

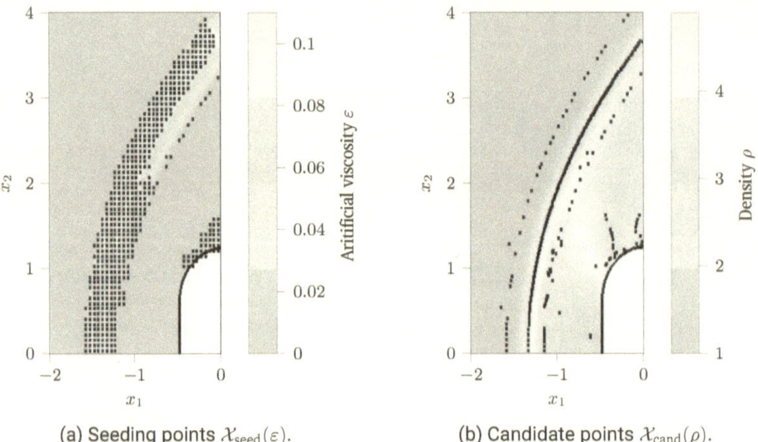

(a) Seeding points $\mathcal{X}_{\mathrm{seed}}(\varepsilon)$.

(b) Candidate points $\mathcal{X}_{\mathrm{cand}}(\rho)$.

Figure 6.8: Shock level-set reconstruction. Point-based search of candidate points $\mathcal{X}_{\mathrm{cand}}(\rho)$ to find the inflection points $\mathcal{X}_{\mathrm{infl}}(\rho)$ of the density field ρ following Algorithm 6.1. Non-converged points have been removed. (a) We seed points $\mathbf{x}_{\mathrm{seed}} \in \mathcal{X}_{\mathrm{seed}}(\varepsilon)$ at the center of cells $K_j \in \mathcal{K}_h^{\mathrm{X}} \subset \Omega_{\mathrm{f}}$ where artificial viscosity is active ($\varepsilon_j > 0$). (b) The auxiliary method FINDCANDIDATEPOINT(...) delivers possible candidates $\mathcal{X}_{\mathrm{cand}}(\rho)$ for the inflection points $\mathcal{X}_{\mathrm{infl}}(\rho)$. Only the upper half of the domain is plotted for better visibility.

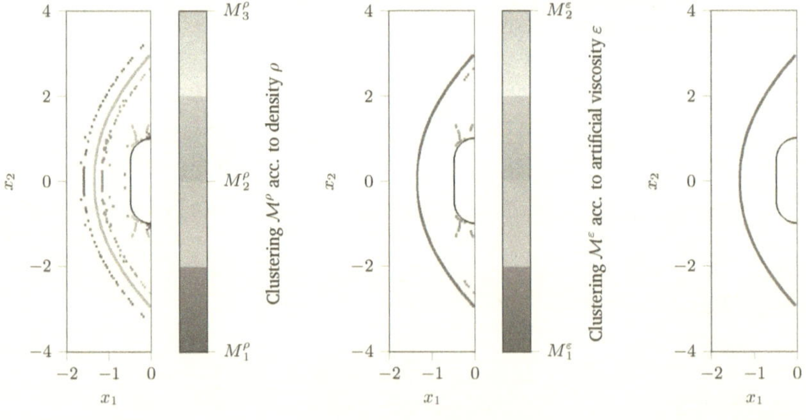

(a) Density-based clustering $\mathcal{M}^\rho$.

(b) Artificial viscosity-based clustering $\mathcal{M}^\varepsilon$.

(c) Inflection points $\mathcal{X}_{\mathrm{infl}}(\rho)$.

Figure 6.9: Shock level-set reconstruction. (a) and (b) We cluster the candidate points $\mathcal{X}_{\mathrm{cand}}(\rho)$ according to their density values ρ and subsequently according to their artificial viscosity ε, see also Figure 6.8b. (c) Finally, we sort out outliers and points around the blunt body to obtain the inflection points $\mathcal{X}_{\mathrm{infl}}(\rho)$.

see Figure 6.9a. We look for the cluster M_m^ρ whose mean value $\overline{\rho}_m = \overline{\rho}(M_m^\rho)$ is closest to the average value of the jump across the shock $\overline{\rho}_s = (\rho_{\text{pre}} + \rho_{\text{post}})/2$ by evaluating the expression

$$\underset{m}{\arg\min} \left| \overline{\rho}_m - \underbrace{\frac{\rho_{\text{pre}} + \rho_{\text{post}}}{2}}_{\overline{\rho}_s} \right| , \qquad m = 1, \ldots, M . \tag{6.17}$$

Here, the cluster $M_{m=2}^\rho$ is selected according to Equation (6.17). The values ρ_{pre} and ρ_{post} can be directly obtained from the numerical simulation. The points $\mathbf{x} \in \{M_1^\rho, M_3^\rho\}$ are no longer considered. All selected points $\mathbf{x} \in M_2^\rho$ are then grouped according to their artificial viscosity values $\varepsilon(\mathbf{x})$, resulting in the clustering $\mathcal{M}^\varepsilon = \{M_1^\varepsilon, M_2^\varepsilon\}$, see Figure 6.9b. We select all points $\mathbf{x} \in M_2^\varepsilon$ and remove outliers and points which are close to the blunt body and the domain boundary. Figure 6.9c shows the remaining points $\mathbf{x} \in M_2^{\varepsilon*}$ which define the inflection points

$$\mathcal{X}_{\text{infl}}(\rho) = M_2^{\varepsilon*} \subset M_2^\varepsilon \subset M_2^\rho \subset \mathcal{X}_{\text{cand}} . \tag{6.18}$$

Shock level-set function and filtering We consider the previously computed inflection points $\mathcal{X}_{\text{infl}}(\rho)$ and reconstruct a shock level-set function $\varphi_s(\mathbf{x})$ according to

$$|\varphi_s(\mathbf{x})| = \min_{\mathbf{x}_{\text{infl}} \in \mathcal{X}_{\text{infl}}} |\mathbf{x} - \mathbf{x}_{\text{infl}}| , \tag{6.19a}$$

$$\text{sign}(\varphi_s(\mathbf{x})) = \text{sign}(\rho(\mathbf{x}) - \rho(\mathcal{X}_{\text{infl}})) . \tag{6.19b}$$

Based on this equation, we introduce an additional partitioning of the computation domain Ω, compare Equation (3.65), as

$$\mathfrak{L}(\varphi_s) = \{\mathbf{x} \in \Omega : \varphi_s(\mathbf{x}) < 0\} , \tag{6.20a}$$

$$\mathfrak{I}_s(\varphi_s) = \{\mathbf{x} \in \Omega : \varphi_s(\mathbf{x}) = 0\} , \tag{6.20b}$$

$$\mathfrak{R}(\varphi_s) = \{\mathbf{x} \in \Omega : \varphi_s(\mathbf{x}) > 0\} , \tag{6.20c}$$

where $\mathfrak{I}_s(\varphi_s) := \overline{\mathfrak{L}(\varphi_s)} \cap \overline{\mathfrak{R}(\varphi_s)}$ and $\Omega = \mathfrak{L}(\varphi_s) \cup \mathfrak{I}_s(\varphi_s) \cup \mathfrak{R}(\varphi_s)$ hold. The species $\mathfrak{L}(\varphi_s)$ and $\mathfrak{R}(\varphi_s)$ denote the pre-shock and the post-shock region, respectively. In particular, $\mathfrak{I}_s(\varphi_s)$ represents a sharp interface description of the shock front.

In general, the projection $\varphi_s^{\text{DG}}(\mathbf{x}) := \text{Proj}_{\mathbb{P}_P^X(\mathcal{K}_h)}(\varphi_s(\mathbf{x}))$ of the shock level-set function $\varphi_s(\mathbf{x})$ onto the XDG space $\mathbb{P}_P^X(\mathcal{K}_h)$ is discontinuous. As a solution, we apply patch-recovery filters and a continuity projection denoted by the general expressions Proj_{prc} and $\text{Proj}_{\text{cont}}$, respectively, in order to stabilize the computation and to obtain a continuous approximation

$$\varphi_s^{\text{DG},C^0}(\mathbf{x}) := \text{Proj}_{\text{cont}} \left(\text{Proj}_{\text{prc}}(\varphi_s^{\text{DG}}(\mathbf{x})) \right) \in \mathbb{P}_P^X(\mathcal{K}_h) \cap C^0 . \tag{6.21}$$

The reader is referred to Sections 6.3.2 and 6.3.3 for details about the applied patch-recovery filters Proj_{prc} and the continuity projection $\text{Proj}_{\text{cont}}$, respectively. It is sufficient to apply the continuity projection only to the narrow band around the shock interface $\mathfrak{I}_s^{\text{DG}}$, since we are only interested in a sharp and continuous description of the shock front. Thus, the shock level-set function (6.21) may be nearly arbitrarily chosen in the remaining parts of the computational

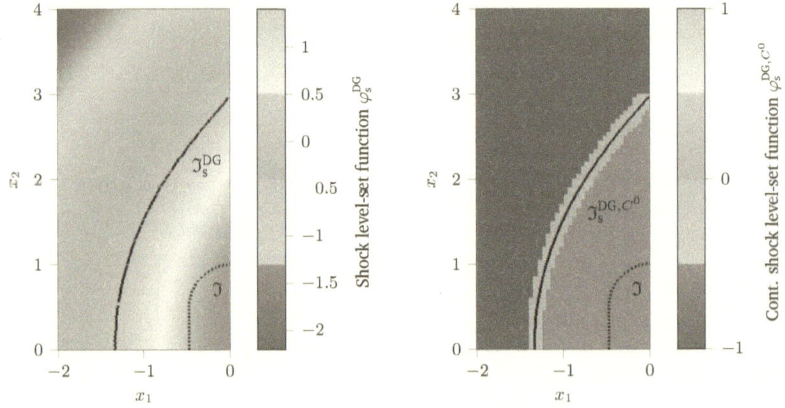

(a) Initially reconstructed, discontinuous shock interface $\mathfrak{I}_s^{\mathrm{DG}}$.

(b) Optimized, continuous shock interface $\mathfrak{I}_s^{\mathrm{DG},C^0}$.

Figure 6.10: Shock level-set reconstruction. (a) We reconstruct a shock level-set function $\varphi_s^{\mathrm{DG}}(\mathbf{x}) \in \mathbb{P}_P^{\mathrm{X}}(\mathcal{K}_h)$ based on the inflection points $\mathcal{X}_{\mathrm{infl}}(\rho)$. (b) Subsequently, we compute a continuous version $\varphi_s^{\mathrm{DG},C^0}(\mathbf{x}) \in \mathbb{P}_P^{\mathrm{X}}(\mathcal{K}_h) \cap C^0$ by using patch-recovery filters and a continuity projection, which is only applied to the narrow band (6.11) around the shock interface $\mathfrak{I}_s^{\mathrm{DG}}$. The dotted interface $\mathfrak{I}$ represents the blunt body. Only the upper half of the domain is plotted for better visibility.

domain, as there only the sign matters. In Figure 6.10, we show the level-set functions $\varphi_\mathrm{s}^\mathrm{DG}(\mathbf{x})$ and $\varphi_\mathrm{s}^{\mathrm{DG},C^0}(\mathbf{x})$ with their interfaces $\mathfrak{I}_\mathrm{s}^\mathrm{DG}$ and $\mathfrak{I}_\mathrm{s}^{\mathrm{DG},C^0}$, respectively, as well as the interface $\mathfrak{I}$ describing the blunt body. The initially reconstructed shock interface $\mathfrak{I}_\mathrm{s}^\mathrm{DG}$ is discontinuous as shown in Figure 6.10a. We apply the projections (6.21) in order to obtain a smooth and continuous version $\mathfrak{I}_\mathrm{s}^{\mathrm{DG},C^0}$, see Figure 6.10b. In this particular case, the values in the so-called *far field* outside of the narrow band $\mathcal{K}_h^{cc,1}$ ($\mathcal{K}_h^{\mathrm{far}} = \mathcal{K}_h^\mathrm{X} \setminus \mathcal{K}_h^{cc,1}$) are set to -1 and $+1$, respectively.

6.4 Sub-Cell Accurate Correction of the Shock Position[17]

In this section, we present a novel algorithm for the correction of the shock interface position inside a cut background cell. We start from a reconstructed shock interface $\mathfrak{I}_\mathrm{s}$, which can be obtained by the technique presented in Section 6.3. Subsequently, the position $x_{\mathfrak{I}_\mathrm{s}}$ of the shock interface is iteratively corrected until it coincides with the exact shock position x_s. In this section, the discussion is limited to a one-dimensional (pseudo-two-dimensional) setting so that the shock interface consists of a single point ($\mathfrak{I}_\mathrm{s} = \{x_{\mathfrak{I}_\mathrm{s}}\}$). In particular, our novel algorithm makes use of an implicit pseudo-time-stepping procedure, which iteratively corrects the current position of the shock interface $x_{\mathfrak{I}_\mathrm{s}}^l$ by means of a suitable cell-local indicator such that $\lim_{l\to\infty} x_{\mathfrak{I}_\mathrm{s}}^l \to x_\mathrm{s}$. All indicators are derived from the local solution in the cut-cells. After the implicit pseudo-time-stepping procedure has terminated, the entire flow field can be advanced in time as done in a standard XDG computation.

This section is structured as follows: We introduce the methodology of implicit time-integration schemes in Section 6.4.1 and derive indicators for the correction of the shock interface position in Section 6.4.2. Both approaches are essential parts of the implicit pseudo-time-stepping procedure. A proof of concept is presented for the test case of a pseudo-two-dimensional stationary normal shock wave in Section 6.4.3.

6.4.1 Implicit Time-Integration Schemes

The temporal integral of the integrated local weak form (4.6) in a cell $K_j \in \mathcal{K}_h^\mathrm{X}$

$$\tilde{\psi}_j(t_{n+1}) = \tilde{\psi}_j(t_n) - \int_{t=t_n}^{t_{n+1}} \mathbf{M}_j^{-1} \mathbf{Op}_j(t, \psi_h(t)) \, \mathrm{d}t \tag{6.22}$$

may be discretized by implicit time-integration schemes. In Equation (6.22), $\tilde{\psi}_j(t_{n+1})$ and $\tilde{\psi}_j(t_n)$ are the unknown and known cell-local coefficient vectors, respectively. The discrete solution on the entire grid is denoted by $\psi_h(t)$. In contrast to the explicit time-integration schemes, which have been introduced for a general DG method in Section 4.1, the inverse of the cell-local mass matrix $\mathbf{M}_j^{-1}$ is present, since the applied XDG method is based on a cut-cell grid $\mathcal{K}_h^\mathrm{X}$, where $\mathbf{M}_j \neq \mathbf{I}$ in general. Implicit time-integration schemes approximate the temporal integral in Equation (6.22) by taking the operator evaluation $\mathbf{Op}_j(t_{n+1}, \psi_h(t_{n+1}))$ at a new

---

time-level t_{n+1} into account. Thus, they require the solution of a (nonlinear) equation system. In this context, the simplest linear multi-step method is the *implicit Euler* scheme

$$\tilde{\psi}_j(t_{n+1}) = \tilde{\psi}_j(t_n) - \Delta t\, \mathbf{M}_j^{-1} \mathbf{Op}_j(t_{n+1}, \psi_h(t_{n+1})) \,, \tag{6.23}$$

which can be seen as a member of the backwards differentiation formula (BDF) schemes. These schemes are A-stable for an order of $Q < 3$ (Dahlquist, 1963) and offer greater stability regions than their explicit equivalent, the linear multi-step Adams-Bashforth (AB) schemes, see also Section 4.1.2. In particular, A-stability means that the time-step size Δt can be chosen arbitrarily large to obtain a stable solution. For a brief discussion of the main differences to explicit time-integration schemes, the reader is referred to Section 4.1. We recommend the work by Deville et al. (2002) for a general introduction.

Solution of nonlinear equation systems We write the implicit Euler scheme (6.23) in a global version by skipping the cell index j as

$$\tilde{\psi}_{n+1} = \tilde{\psi}_n - \Delta t\, \mathbf{M}^{-1} \mathbf{Op}(t_{n+1}, \psi_h(t_{n+1})) \,, \tag{6.24}$$

where we use the notations $\tilde{\psi}_{n+1} = \tilde{\psi}(t_{n+1})$ and $\tilde{\psi}_n = \tilde{\psi}(t_n)$. The initial solution $\tilde{\psi}_0 = \tilde{\psi}(t_0)$ is given by the projection of the initial conditions of a test case. Equation (6.24) requires the solution of a nonlinear equation system in the case of the Euler equations (2.5). In this work, we apply Newton's method to solve the nonlinear system in each time-step (Kelley, 1995, Chapter 5). For that, we rewrite Equation (6.24) in the form

$$\mathbf{F}(\tilde{\psi}_{n+1}) := \tilde{\psi}_{n+1} - \tilde{\psi}_n + \Delta t\, \mathbf{M}^{-1} \mathbf{Op}(t_{n+1}, \psi_h(t_{n+1})) = 0 \,. \tag{6.25}$$

In general, nonlinear iterative methods seek for a solution of

$$\mathbf{F}(\tilde{\psi}) = 0 \,. \tag{6.26}$$

If the components of the vector $\mathbf{F}(\tilde{\psi})$ are differentiable, the Jacobian matrix is defined as

$$\mathbf{F}'(\tilde{\psi}) = \left(\mathbf{F}'(\tilde{\psi})\right)_{i,j} = \frac{\partial F_i}{\partial \tilde{\psi}_j}(\tilde{\psi}) \,. \tag{6.27}$$

Equation (6.25) may be solved by Newton's method, which calculates a new iterate $\tilde{\psi}_{n+1}^{k+1}$ based on the current iterate $\tilde{\psi}_{n+1}^{k}$ via

$$\tilde{\psi}_{n+1}^{k+1} = \tilde{\psi}_{n+1}^{k} - \left(\mathbf{F}'(\tilde{\psi}_{n+1}^{k})\right)^{-1} \mathbf{F}(\tilde{\psi}_{n+1}^{k}) \,. \tag{6.28}$$

Note that the index k denotes the current iteration of the nonlinear solver and n denotes the current time-step. A starting guess for $\tilde{\psi}_{n+1}^{1}$ is given by the numerical solution from the last time-step $\tilde{\psi}_n$ or the initial condition $\tilde{\psi}_0$, respectively.

In the *BoSSS* framework, the Jacobian matrix (6.27) can be approximated by a finite difference approach where every j-th column is calculated by

$$\left(\mathbf{F}'(\tilde{\psi})\right)_{-,j} \approx \frac{1}{\sqrt{\varepsilon_{\mathrm{d}}}} \left(\mathbf{F}(\tilde{\psi} + \sqrt{\varepsilon_{\mathrm{d}}}\mathbf{e}_j) - \mathbf{F}(\tilde{\psi})\right) \,. \tag{6.29}$$

Here, $\varepsilon_d = 1.11 \cdot 10^{-16}$ denotes the machine accuracy in terms of double precision (Kelley, 1995, Section 5.4). The perturbation $\sqrt{\varepsilon_d}\mathbf{e}_j$ of the j-th degree of freedom, which we assume to be located in cell K_l, only affects the operator evaluation in cell K_l and in all of its neighbors. Therefore, by exploiting the locality of the XDG-operator, the assembly of the finite difference approximation to the Jacobian matrix can be optimized so that the number of required evaluations scales only with the polynomial degree, but not with the total number of cells. If analytical derivatives are available, the *BoSSS* framework offers faster approaches than the finite difference approximation. This topic is beyond the scope of this work.

We apply a line-search method to improve the global convergence of Newton's method (6.28). We use the fact that a globally convergent iterative method converges for every initial solution or fails after a small number of steps (Kelley, 1995, Chapter 8). Consequently, the so-called condition for *sufficient decrease* has to be fulfilled in each iteration. We first calculate a Newton direction

$$\tilde{\mathbf{d}} = - \left(\mathbf{F}'(\tilde{\psi}^k) \right)^{-1} \mathbf{F}(\tilde{\psi}^k) . \tag{6.30}$$

Second, we calculate test steps $\tilde{s} = \tilde{\lambda}\tilde{\mathbf{d}}$ with $\tilde{\lambda} = 2^m$ for $m \geq 0$ until $\mathbf{F}(\tilde{\psi}^k + \tilde{s})$ fulfills the condition for *sufficient decrease*

$$\left| \mathbf{F}(\tilde{\psi}^k + \tilde{\lambda}\tilde{\mathbf{d}}) \right| < (1 - \tilde{\alpha}\tilde{\lambda}) \left| \mathbf{F}(\tilde{\psi}^k) \right| , \tag{6.31}$$

where the parameter $\tilde{\alpha} \in (0, 1)$ is a small, positive number chosen so that Equation (6.31) is simple to satisfy (Kelley, 1995, Chapter 8).

Every iteration of Newton's method requires the solution of a linear equation system. If the problem size is small enough, direct solvers, such as the libraries Parallel Direct Solver (PARDISO) (Schenk et al., 1999) and Multifrontal Massively Parallel Sparse Direct Solver (MUMPS) (Amestoy et al., 2001; Amestoy et al., 2006), can still be employed efficiently. In this work, we apply the MUMPS framework. Before solving the linear system, the condition number of the system matrix is reduced by the cell-agglomeration technique, which removes small and ill-shaped cut-cells from the original cut-cell grid, see Section 3.3.2.

6.4.2 Indicators

If the shock interface position does not coincide with the exact shock position ($x_{\mathfrak{I}_s} \neq x_s$), the polynomial solution starts to oscillate in the vicinity of a discontinuity. These oscillations usually spread all over the computational domain and render the numerical solution unstable. Therefore, a correction of the shock interface position $x_{\mathfrak{I}_s}$ is inherently necessary for a stable high-order XDG method in the context of supersonic compressible flow.

In this section, we derive several indicators in order to correct the position of the reconstructed shock front in a cut background cell. In particular, we introduce an indicator which is based on a $P0$-projection of the local solution and two indicators which rely on the deviation from the normal shock relations.

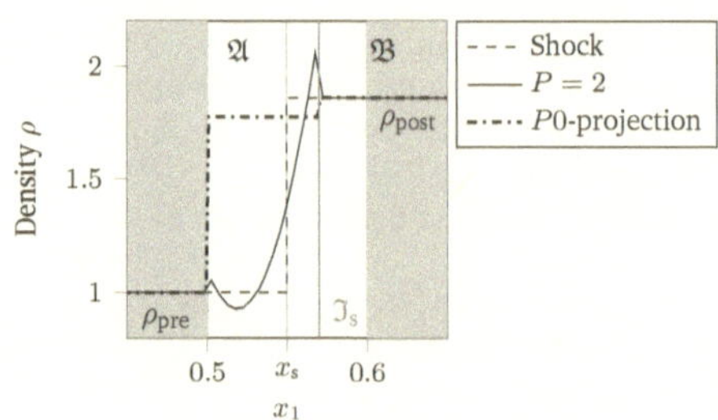

Figure 6.11: $P0$-projection of the local solution in a cut background cell $K_j = [0.5, 0.6]$. A stationary shock wave with a Mach number of $M_s = 1.5$ is located at $x_s = 0.55$. The solution oscillates, since the shock interface position does not coincide with the exact shock position ($x_{\mathfrak{J}_s} \neq x_s$). The $P0$-projection considers only the zeroth-order mode in species $\mathfrak{A}$ and $\mathfrak{B}$. The exact density values ρ_{pre} and ρ_{post} can be extracted from the near band $\mathcal{K}_h^{\mathrm{near}} := \mathcal{K}_h^{cc,1} \setminus \mathcal{K}_h^{cc,0}$ (shown in gray) for later usage (adapted from Geisenhofer et al., 2020, Figure 1).

Indicator based on a $P0$-projection We consider the local polynomial solution in a species $\mathfrak{s}$ in a cut-cell $K_{j,\mathfrak{s}} \in \mathcal{K}_h^X$ with $K_{j,\mathfrak{s}} \neq \emptyset$ denoting it by

$$\psi_{j,\mathfrak{s}}(\mathbf{x}, t) := \psi|_{K_{j,\mathfrak{s}}} \in \mathbb{P}_P^X(\{K_{j,\mathfrak{s}}\}) . \tag{6.32}$$

If $x_{\mathfrak{J}_s} \neq x_s$, it is hardly possible to extract reasonable physical information from the oscillating numerical solution. Therefore, we remove the high-order modes from the solution and extract solely the zeroth-order mode

$$\psi_{j,\mathfrak{s}}^{P0}(\mathbf{x}, t) = \Pi^{P0}(\psi_{j,\mathfrak{s}}) , \tag{6.33}$$

by using the projection operator

$$\begin{aligned}
\Pi^{P0} : \mathbb{P}_P^X(\{K_{j,\mathfrak{s}}\}) &\to \mathbb{P}_{P=0}^X(\{K_{j,\mathfrak{s}}\}) , \\
\psi &\mapsto \psi^{P0} ,
\end{aligned} \tag{6.34}$$

with the essential property $\langle \psi - \psi^{P0}, \vartheta \rangle = 0$, $\forall \vartheta \in \mathbb{P}_{P=0}^X(\{K_{j,\mathfrak{s}}\})$. As an example, Figure 6.11 shows the numerical solution for a stationary shock wave with a Mach number of $M_s = 1.5$ located at $x_s = 0.55$ on a cut background cell $K_j = [0.5, 0.6]$ with the species $\mathfrak{A}$ and $\mathfrak{B}$. Since the shock interface position $x_{\mathfrak{J}_s} = 0.57$ does not coincide with the exact shock position $x_s = 0.55$, the high-order solution ($P = 2$) oscillates. Additionally, we show the $P0$-projection of the high-order solution using the projection operator (6.34). Note that in general the projected values in species $\mathfrak{A}$ and $\mathfrak{B}$ are different ($\psi_{j,\mathfrak{A}}^{P0} \neq \psi_{j,\mathfrak{B}}^{P0}$), since there is a separate set of degrees of freedom (DOF) for each species.

Subsequently, we construct an indicator for the correction of the shock position based on the projection operator (6.34). For this purpose, we consider the solution in the so-called *near band*, which is shown in gray in Figure 6.11. The near band $\mathcal{K}_h^{\mathrm{near}} := \mathcal{K}_h^{cc,1} \setminus \mathcal{K}_h^{cc,0}$ is defined as the set of cells consisting of solely the neighboring cells of all cut-cells, compare Equations (6.10) and (6.11). In this test case, the numerical solution equals the exact solutions ρ_{pre} and ρ_{post} inside the near band. In other cases, the width of the near band may be extended. We define the local indicator $\mathcal{I}_j^{P0}(\rho)$ acting on the density ρ in a cut background cell $K_j \in \mathcal{K}_h$ with $K_{j,\mathfrak{s}} \neq \emptyset$ as

$$\mathcal{I}_j^{P0}(\rho) = \mathcal{I}^{P0}(\rho)|_{K_j} := \underbrace{\frac{\rho_{\text{pre}} + \rho_{\text{post}}}{2}}_{\substack{\text{exact solution in} \\ \text{near band } \mathcal{K}_h^{\text{near}}}} - \underbrace{\frac{\rho_{j,\mathfrak{A}}^{P0} + \rho_{j,\mathfrak{B}}^{P0}}{2}}_{\substack{P0\text{-projections} \\ \text{in cut-cells } K_{j,\mathfrak{s}}}} , \tag{6.35}$$

where we compare the average value of the exact solution, which is extracted from the near band, to the average value of the $P0$-projection in both species $\mathfrak{A}$ and $\mathfrak{B}$. The final shifting direction of the shock interface $\mathfrak{I}_{\mathfrak{s}}$ is given by the sign of $\mathcal{I}_j^{P0}(\rho)$ according to

$$\text{if } \mathcal{I}_j^{P0}(\rho) > 0 \quad \Rightarrow \quad \text{shift } \mathfrak{I}_{\mathfrak{s}} \text{ to the right}, \tag{6.36a}$$
$$\text{if } \mathcal{I}_j^{P0}(\rho) < 0 \quad \Rightarrow \quad \text{shift } \mathfrak{I}_{\mathfrak{s}} \text{ to the left}. \tag{6.36b}$$

In other words, a value of $\mathcal{I}_j^{P0}(\rho) < 0$ indicates that the average value of the $P0$-projections is larger than the exact average value of the shock, see Equation (6.35). This means that the shock interface $\mathfrak{I}_{\mathfrak{s}}$ is located too far right compared to the exact shock position in the cut background cell ($x_{\mathfrak{I}_{\mathfrak{s}}} > x_{\mathfrak{s}}$). The new shock interface position $x_{\mathfrak{I}_{\mathfrak{s}}}^{l+1}$ is obtained by a simple bisection algorithm based on the two previous positions $x_{\mathfrak{I}_{\mathfrak{s}}}^{l}$ and $x_{\mathfrak{I}_{\mathfrak{s}}}^{l-1}$ or during the first iterations also based on the cell boundaries $x_1(\partial K_j)$, respectively. For the scenario shown in Figure 6.11, the indicator value is negative ($\mathcal{I}_j^{P0}(\rho) = -0.389$) so that the new shock interface position is obtained by $x_{\mathfrak{I}_{\mathfrak{s}}}^{l+1} = (x_{\mathfrak{I}_{\mathfrak{s}}}^{l} + x_1[\partial K_{j,\text{left}}])/2 = (0.57 + 0.5)/2 = 0.535$.

Indicators based on the normal shock relations We repeat the normal shock relations (2.30)

$$\rho_{\text{pre}} u_{1,\text{pre}} = \rho_{\text{post}} u_{1,\text{post}} \qquad \text{(continuity equation)}, \qquad \text{(2.30a, repeated)}$$
$$p_{\text{pre}} + \rho_{\text{pre}} u_{1,\text{pre}}^2 = p_{\text{post}} + \rho_{\text{post}} u_{1,\text{post}}^2 \qquad \text{(momentum equation)}, \qquad \text{(2.30b, repeated)}$$

for the definition of two additional indicators. If a numerical scheme resolves a shock in a sharp manner, the normal shock relations have to be fulfilled up to the accuracy of the numerical scheme. A deviation at the shock interface $\mathfrak{I}_{\mathfrak{s}}$ motivates the definition of the indicators $\mathcal{I}^{\rho}(\mathbf{U})$ and $\mathcal{I}^m(\mathbf{U})$, which rely on Equations (2.30a, repeated) and (2.30b, repeated), respectively. We define the indicator $\mathcal{I}_j^{\rho}(\mathbf{U})$ in a cut background cell $K_j \in \mathcal{K}_h$ with $K_{j,\mathfrak{s}} \neq \emptyset$ as

$$\begin{aligned} \mathcal{I}_j^{\rho}(\mathbf{U}) = \mathcal{I}^{\rho}(\mathbf{U})|_{K_j} &:= \frac{1}{\partial K_j} \oint_{\mathfrak{I}} \rho_{j,\mathfrak{A}} u_{1,j,\mathfrak{A}} - \rho_{j,\mathfrak{B}} u_{1,j,\mathfrak{B}} \, \mathrm{d}S \\ &= \frac{1}{\partial K_j} \oint_{\mathfrak{I}} [\![\rho_j u_{1,j}]\!] \, \mathrm{d}S, \end{aligned} \tag{6.38}$$

and the indicator $\mathcal{I}_j^m(\mathbf{U})$ as

$$\begin{aligned} \mathcal{I}_j^m(\mathbf{U}) = \mathcal{I}^m(\mathbf{U})|_{K_j} &:= \frac{1}{\partial K_j} \oint_{\mathfrak{I}} p_{j,\mathfrak{A}} + \rho_{j,\mathfrak{A}} u_{1,j,\mathfrak{A}}^2 - (p_{j,\mathfrak{B}} + \rho_{j,\mathfrak{B}} u_{1,j,\mathfrak{B}}^2) \, \mathrm{d}S \\ &= \frac{1}{\partial K_j} \oint_{\mathfrak{I}} [\![p_j + \rho_j u_{1,j}^2]\!] \, \mathrm{d}S. \end{aligned} \tag{6.39}$$

The corrected shock interface position $x_{\mathfrak{I}_{\mathfrak{s}}}^{l+1}$ is obtained by the bisection algorithm as presented for the indicator $\mathcal{I}^{P0}(\rho)$.

6.4.3 Proof of Concept

In the beginning of Section 6.3, we have considered a supersonic blunt-body problem computed with a DG IBM using a two-step shock-capturing strategy based on artificial viscosity. By showing the limitations of this approach, we have motivated the application of a high-order XDG method. Furthermore, we have derived the necessary steps to regain the high-order convergence rate. In this section, we present the results of a novel implicit pseudo-time-stepping procedure, which is based on the methodology introduced in Sections 6.4.1 and 6.4.2. In particular, this procedure incorporates a sub-cell accurate correction of the reconstructed shock interface so that it coincides with the exact shock position.

As a test case, we investigate a pseudo-two-dimensional normal shock wave of strength $M_s = 1.5$ in a pseudo-stationary state by fixing the frame of reference to the shock wave. The shock Mach number M_s determines the physical conditions in the pre- and post-shock state. An introduction to unsteady wave motion and in particular to the different frames of reference has been given in Section 2.2.3. We choose a computational domain of $\Omega = [0, 1] \times [0, 0.3]$ consisting of 10×3 cells with $h = 0.1$. Adiabatic slip-wall boundary conditions are applied on the top and bottom boundary. The exact pre- and post-shock states are prescribed at the left and right boundary, respectively. The shock wave is located at $x_s = 0.55$. The pre-shock initial conditions are given by

$$
(\rho_{\mathrm{pre}}, u_{1,\mathrm{pre}}, u_{2,\mathrm{pre}}, p_{\mathrm{pre}})^{\mathsf{T}} = \left(1, \sqrt{\gamma \frac{p_{\mathrm{pre}}}{\rho_{\mathrm{pre}}}} \, M_s, \, 0, \, 1\right)^{\mathsf{T}}. \qquad (5.38, \text{repeated})
$$

The post-shock conditions are calculated according to Equation (5.39), see also Section 5.5.2. We set the shock-capturing parameters to $S_0 = 1.0 \cdot 10^{-3}$, $\varepsilon_0 = 1.0$, and $\kappa = 0.5$ and apply the explicit Euler scheme (4.7) for advancing the solution in time until $t_{\mathrm{end}} = 2.0$. Before we derive the implicit-pseudo-time-stepping procedure, we briefly summarize the road map to it.

Limitations of the two-step shock-capturing strategy based on artificial viscosity In a high-order DG method, the polynomial solution immediately starts to oscillate in the vicinity of a large gradient or a discontinuity as soon as the shock-capturing scheme is deactivated. Figure 6.12 shows an illustrating example. We show the solutions for activated and deactivated shock-capturing for a polynomial degree of $P = 4$. The sharpness of the shock front can be optimized by more sophisticated variants, but will always be limited by the application of nonphysical artificial viscosity (Barter and Darmofal, 2007; Klöckner et al., 2011; Lv et al., 2016). Consequently, the smeared-out results with a thickness of $O(h/P)$ are optimal to a certain extent. Even for solutions where the steady state is already reached, the deactivation of the shock-capturing scheme will cause oscillations in the vicinity of already smeared-out discontinuities due to the self-steepening character of the Euler equations (2.5).

Challenges for explicit time-integration schemes In general, reaching the steady-state solution of hyperbolic partial differential equations (PDEs) may be challenging when applying explicit time-integration schemes. For example, this is problematic due to the formation of nonphysical waves which are caused by initial or boundary conditions, see also Section 5.5.4. In such cases, the steady state can only be reached if the nonphysical waves are damped out by numerical dissipation or propagate outside of the computational domain. Furthermore,

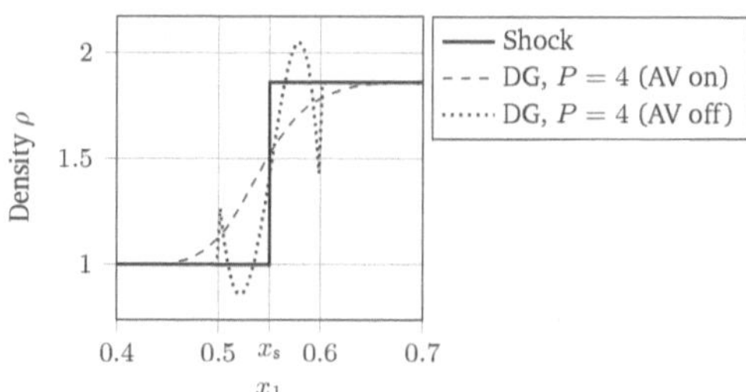

Figure 6.12: Stationary shock wave computed with a DG method. The converged steady-state solution is obtained by the explicit Euler scheme at $t_{\mathrm{end}} = 2.0$ (AV on). As soon as the two-step shock-capturing strategy based on artificial viscosity (AV) is deactivated (AV off), oscillations arise in the vicinity of the shock.

explicit time-step restrictions may limit the application from a practical point of view for large simulation end times. However, we note that this effect strongly depends on the underlying physics of the test case.

Extended discontinuous Galerkin method as a remedy An essential step of any DG method is the solution of local Riemann problems. In the XDG method, an additional Riemann problem is solved at the interface between different species or the pre- and post-shock state, respectively. In Section 6.3, we have a derived a reconstruction procedure of a level-set function, whose zero iso-contour represents a *cell-accurate* description of the shock front. For complex shock shapes such as a bow shock, a correction mechanism is required in order to obtain a *sub-cell accurate* description inside a cut background cell.

Implicit pseudo-time-stepping procedure In the following, we present a novel implicit pseudo-time-stepping procedure for determining the exact shock interface position in a cut background cell based on an initial cell-accurate guess. We consider the stationary shock wave as described in the beginning of this section on the computational domain $\Omega = [0, 1] \times [0, 0.3]$ with a coarse grid consisting of 10×3 cells with $h = 0.1$. A zoom-in is shown in Figure 6.13. The exact position of the shock wave is set to $x_s = 0.55$. In order to compute a sufficiently smooth initial condition, we define a smoothed Heaviside function $\tilde{H}(\mathbf{x})$ as

$$\tilde{H}(\mathbf{x}) = 0.5 \left[\tanh\left(\frac{|\mathbf{x} - \mathbf{x}_s|}{\tilde{C}\, h/\max(0, P)} \right) + 1.0 \right], \tag{6.40}$$

where $\mathbf{x}$ is an arbitrary point in space, $\mathbf{x}_s$ is the position of the shock front, and $\tilde{C}$ is a user-defined factor in order to control the strength of the smoothing, which we set to $\tilde{C} = 1.0$. Subsequently, the smoothed initial condition of a scalar quantity $\psi_0(\mathbf{x}) = \psi(\mathbf{x}, t_0)$ is calculated with

$$\psi_0(\mathbf{x}) = \psi_{\mathrm{pre}} - \tilde{H}(\mathbf{x})(\psi_{\mathrm{pre}} - \psi_{\mathrm{post}}). \tag{6.41}$$

In this test case, we apply the initial smoothing to all variables of the state vector $\mathbf{U} = (\rho, \rho u_1, \rho u_2, \rho E)^\mathsf{T}$, see also Equation (2.6). This mimics the solution of a standard DG method

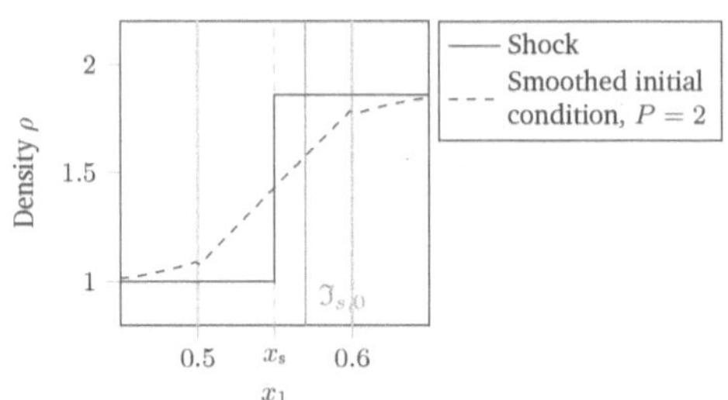

Figure 6.13: Initial configuration for determining the sub-cell accurate position of a stationary shock wave in an XDG method. The smoothed initial condition mimics the solution of standard DG method with shock-capturing. The shock is located at $x_s = 0.55$. The initial shock interface position is assumed to be at $x_{\mathfrak{I}_s,0} = x_{\mathfrak{I}_s}(t_0) = 0.57$ (adapted from Geisenhofer et al., 2020, Figure 2).

with a shock-capturing strategy. We decide to use Equation (6.41), since the conditioning of the nonlinear equation system and, thus, the convergence of Newton's method (6.28) strongly depend on the initial guess. We could not obtain a convergent Newton's method if the initial condition was too far away from the exact solution. We abstain from investigating this issue even further, since preconditioning techniques and advanced implicit methods in the context of the Euler equations (2.5) are beyond the scope of this work. The reader is referred to the works by Dolejší and Feistauer (2004), Fidkowski et al. (2005), and Shahbazi et al. (2009) for further details. As a remedy, we solve the underlying steady-state problem with an unsteady implicit Euler scheme (6.24) with finite time-step sizes Δt, which mathematically corresponds to under-relaxation. However, we do not expect any limitation in using a standard DG computation with shock-capturing as an input for the XDG computation.

In order to show the robustness of the novel implicit pseudo-time-stepping procedure, we assume an initial shock interface position of $x_{\mathfrak{I}_s,0} = x_{\mathfrak{I}_s}(t_0) = 0.57$, which is sufficiently far away from the exact shock position $x_s = 0.55$. This is equivalent to a level-set function of the form

$$\varphi(x_1) = x_1 - x_{\mathfrak{I}_s,0} = x_1 - 0.57 \,. \tag{6.42}$$

The initial shock interface position $x_{\mathfrak{I}_s,0}$ could have also been obtained by the level reconstruction procedure presented in Section 6.3. Note that in every correction step of the implicit pseudo-time-stepping procedure, the shock interface $\mathfrak{I}_s$ is fixed in space. The procedure terminates if the shock interface position $x_{\mathfrak{I}_s}$ coincides with the exact shock position x_s.

Starting from an initial numerical solution $\mathbf{U}_0(x_1) = \mathbf{U}(x_1, t_0)$ and an initial shock interface position $x_{\mathfrak{I}_s,0}$, we set up an iterative procedure $t_n \to t_{n+1}$, where we discretize the Euler equations (2.5) by the presented XDG method in space and by the unsteady implicit Euler scheme (6.24) in time. We prescribe finite time-step sizes of $\Delta t = 0.1$ in order to improve the convergence of the nonlinear problem to the steady-state solution. We apply a Riemann solver based on Godunov's method (Toro, 2009, Section 4.9) for the spatial discretization of the convective fluxes in contrast to the other test cases of this work, where we use the Harten-Lax-van Leer-Compact (HLLC) flux (Toro et al., 1994), see also Section 3.2.3. It is worth mentioning that well-known wave speed estimates applied in the HLLC flux may fail in cases where high-speed flows impinge on solid stationary walls (Toro, 2009, Chapter 10.9). Additionally, we could confirm this fact in the case of a stationary shock wave, which can be considered as a similar scenario. Thereby, round-off errors determine the sign of the intermediate wave speed (3.45) and, thus, which state in the star region is chosen, see also Figure 3.3. As a result, the HLLC flux does not lead to a robust and reliable result.

Results The results of the implicit pseudo-time-stepping procedure are depicted in Figure 6.14. We show the shock interface positions $x_{\mathfrak{J}_s}$ inside a cut background cell $K_j = [0.5, 0.6]$ in the left column, the unlimited solutions with a polynomial degree of $P = 2$ and the corresponding $P0$-projection in the middle column as well as the values of the indicators $\mathcal{I}^{P0}(\rho)$, $\mathcal{I}^{\rho}(\mathbf{U})$ and $\mathcal{I}^{m}(\mathbf{U})$ in the right column. The results are shown for the pseudo-time-steps #0, #1, #2, and in particular for the converged sharp solution denoted by #last. Here, the shock interface position coincides with the exact shock position ($x_{\mathfrak{J}_s,\text{last}} = x_s = 0.55$). In this test case, only the indicator $\mathcal{I}^{P0}$ delivers the correct shifting direction of the interface. At the pseudo-time-step #1, all indicators deliver the correct shifting direction, whereas at #0 and #2, the indicators $\mathcal{I}^{\rho}$ and $\mathcal{I}^{m}$ based on the normal shock relations deliver the wrong shifting direction, see the right column of Figure 6.14. The new interface positions are obtained by the bisection algorithm presented for the indicator $\mathcal{I}^{P0}$ in Section 6.4.2.

6.5 Summary

In this chapter, we have introduced the concept of shock-fitting for computations of supersonic flows in Sections 6.1 and 6.2. In particular, we have derived a novel reconstruction procedure of a sub-cell accurate sharp representation of the shock front in the context of an XDG method in Sections 6.3 and 6.4. In the following, we summarize the employed methodology for the reconstruction of the shock level-set function. Additionally, we present the main features of the *XDGShock* solver of the *BoSSS* framework, which were implemented during the course of this work.

The computation of supersonic flows containing discontinuous flow phenomena can be realized by means of *shock-capturing* approaches in the context of DG methods, see Chapter 5. On the one hand, these approaches increase the robustness and stability of the numerical method enabling a solution in the first place, but on the other hand, they lead to a loss of global high-order accuracy when adding artificial viscosity. We aim to regain the high-order spatial convergence rate by developing a *shock-fitting* approach in the context of an XDG method. To this end, a sub-cell accurate representation of the sharp shock front is essential. In Section 6.3, we have presented a novel reconstruction procedure of a continuous shock level-set function $\varphi_s^{\text{DG},C^0}(\mathbf{x}) \in \mathbb{P}_P^X(\mathcal{K}_h) \cap C^0$, whose zero iso-contour

$$\mathfrak{J}_s^{\text{DG},C^0} = \left\{ \mathbf{x} \in \Omega : \varphi_s^{\text{DG},C^0}(\mathbf{x}) = 0 \right\} \tag{6.43}$$

describes the shock front, see also Equation (6.21). Figure 6.15 gives an overview of the reconstruction procedure. We consider the inflection points $\mathcal{X}_{\text{infl}}(\rho)$ of the density field ρ as a sufficiently accurate approximation of the exact shock position. For that, we have extracted candidate points $\mathcal{X}_{\text{cand}}(\rho)$ for $\mathcal{X}_{\text{infl}}(\rho)$ from a DG IBM simulation of a supersonic blunt-body

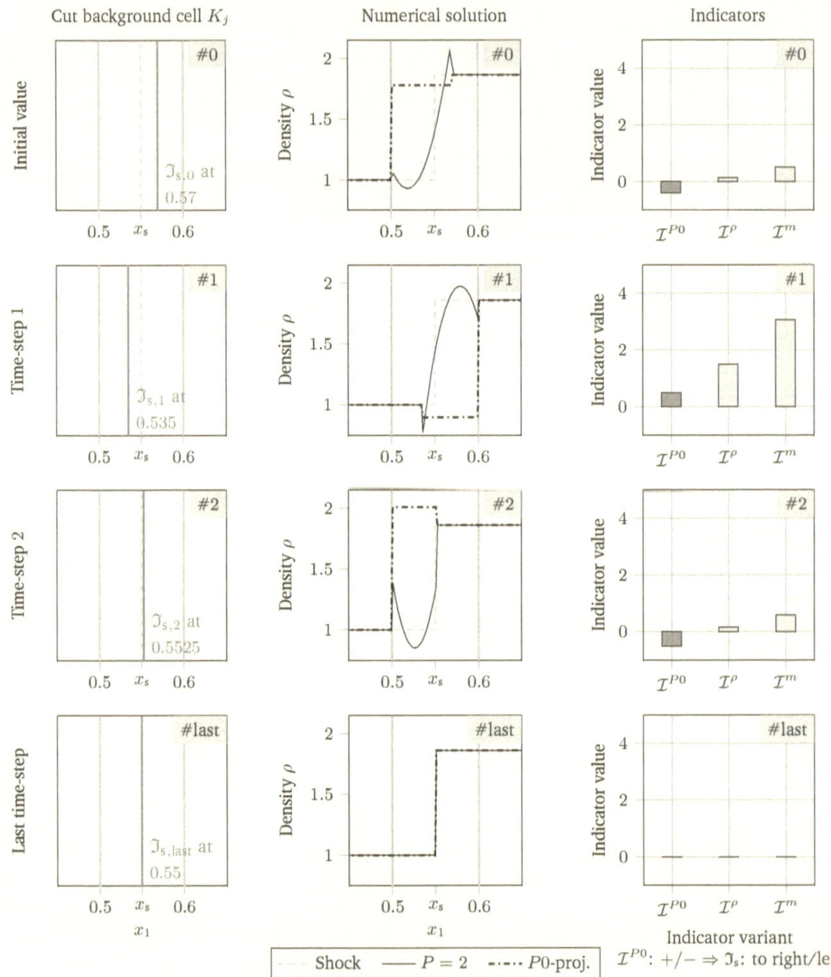

Figure 6.14: Illustration of the implicit pseudo-time-stepping procedure. The position of the shock interface $\mathfrak{I}_s$ converges to the exact shock position $x_s = 0.55$ by using the indicator $\mathcal{I}^{P0}$. Three different indicators are presented: $\mathcal{I}^{P0}(\rho)$ is based on a $P0$-projection of a high-order solution ($P = 2$) of the density field ρ, whereas $\mathcal{I}^\rho(\mathbf{U})$ and $\mathcal{I}^m(\mathbf{U})$ are based on the normal shock relations (2.30a) and (2.30b), respectively. The shock interface position inside the cut background cell $K_j = [0.5, 0.6]$ is iteratively adapted by applying a bisection procedure depending on the sign of $\mathcal{I}^{P0}$. In this test case, only the indicator $\mathcal{I}^{P0}$ delivers the correct shifting direction. The pseudo-time-steps are indicated with #-labels (adapted from Geisenhofer et al., 2020, Figure 3).

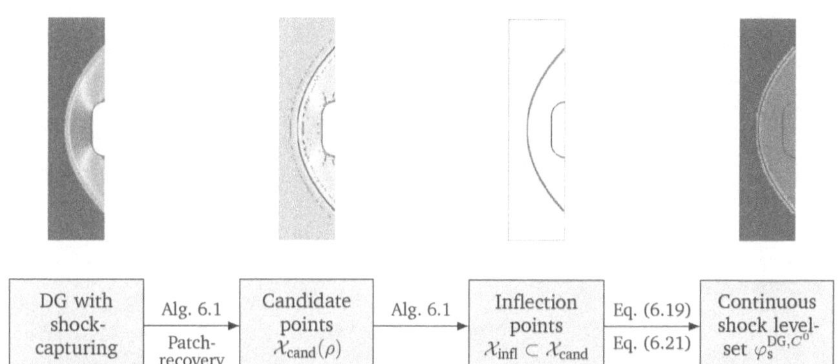

Figure 6.15: Overview of the reconstruction procedure of the shock level-set function.

problem. We have employed the K-means algorithm (Macqueen, 1967) to cluster and sort the candidate points $\mathcal{X}_{\mathrm{cand}}(\rho)$, before we have reconstructed a continuous DG approximation of the shock level-set function from the remaining inflection points $\mathcal{X}_{\mathrm{infl}}(\rho) \subset \mathcal{X}_{\mathrm{cand}}(\rho)$. Additionally, we have applied patch-recovery filters (Kummer and Warburton, 2016) and a continuity projection (Smuda, 2020; Smuda and Kummer, 2020) during the reconstruction procedure. For currently ongoing and future work, the *XDGShock* solver has been verified for the application of two level-set functions in a test case of a subsonic flow over a Gaussian bump in combination with the cell-agglomeration technique, see also Section 3.3.2.

In Section 6.4, we have presented a novel algorithm for the sub-cell accurate adaption of the shock interface position based on the continuous shock level-set function $\varphi_s^{\mathrm{DG},C^0}(\mathbf{x})$. We have developed several indicators for correcting the position of the shock interface inside a cut background cell. The indicator $\mathcal{I}^{P0}(\rho)$ based on a $P0$-projection of the local polynomial solution of the density field performed best in the test case of a pseudo-stationary normal shock wave, see Equation (6.35). The exact shock position has been determined by a newly developed implicit pseudo-time-stepping procedure, for which we have shown a proof of concept in Section 6.4.3. Hereby, we have solved the unsteady two-dimensional Euler equations (2.5) with an under-relaxation-like method, where we have applied an implicit Euler scheme with finite time-step sizes. We have employed Newton's method and the MUMPS framework for solving the nonlinear algebraic equation system without the use of additional preconditioning techniques.

In Table A.2, we state the simulation parameters of the test cases presented in Sections 6.3 and 6.4 for the sake of reproducibility.

XDGShock solver of the *BoSSS* framework   A main goal of this work is the development and implementation of the *XDGShock* solver, which is capable of solving the compressible Euler equations (2.5) with high-order accuracy by using an XDG method. Figure 6.16 gives an overview of the features of the *XDGShock* solver, which extends the *CNS* solver of the *BoSSS* framework. Much effort has been put into the refactoring of the underlying *CNS* solver by porting already implemented features, such as the spatial discretization by numerical fluxes,

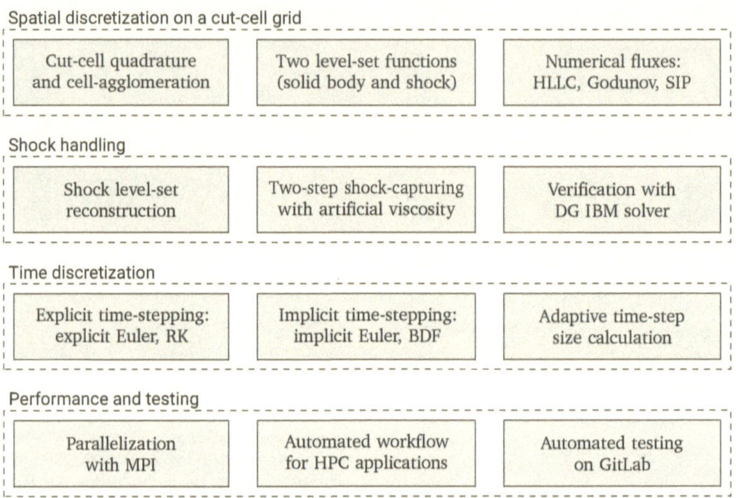

Figure 6.16: Features of the *XDGShock* solver of the *BoSSS* framework, solving the Euler equations for inviscid compressible flow by means of an XDG method.

from *Layer 4 (Application)* to *Layer 3 (Solution)* in order to build a common basis for both solvers, see also Figure 1.1. Furthermore, the *XDGShock* solver has been verified with subsonic and supersonic test cases, which were also used to verify the *CNS* solver, such as the flow over a Gaussian bump, see Section 6.3.1, or an isentropic vortex (Lamb, 1924; Hu, 2006) (not shown in this work).

The main novelties of the *XDGShock* solver are a reconstruction procedure of the shock level-set function and a sub-cell accurate correction of the shock interface position by means of an implicit pseudo-time-stepping procedure, see Sections 6.3 and 6.4, respectively. The *XDGShock* solver enables the application of two level-set functions in combination with the cell-agglomeration technique on a cut-cell grid, see Sections 3.3.2 and 6.3.1, respectively. All newly implemented features can already be used or adapted for parallel simulation runs based on the Message Passing Interface (MPI), aiming at High-Performance Computing (HPC) applications. The submission of computations to small and mid-large computing clusters such as the Hessischer Hochleistungsrechner (HHLR) (*Lichtenberg*) has successfully been automatized. The results obtained by the *XDGShock* solver are reproducible in every build of the *BoSSS* framework due to automated testing on GitLab[18].

7 Conclusion

The conclusion of this work consists of two parts. Section 7.1 summarizes the numerical results of classical benchmarks for supersonic compressible flow, where we have employed a *shock-capturing* strategy and a *shock-fitting* approach in the context of *unfitted* discontinuous Galerkin (DG) methods. Furthermore, it focuses on the novel contributions of this work to the research area of high-order DG methods.

Section 7.2 deals with potential future work and applications. In particular, we show preliminary results of a two-dimensional supersonic blunt-body problem computed with an extended discontinuous Galerkin (XDG) method. The results of this test case represent our current state on the way to a high-order XDG method for high Mach number flows containing discontinuous flow phenomena.

7.1 Summary and Contributions

In this work, we have employed two different types of numerical approaches for the simulation of supersonic compressible flow, namely a *shock-capturing* strategy and a *shock-fitting* approach in the context of a discontinuous Galerkin immersed boundary method (DG IBM) and an extended discontinuous Galerkin (XDG) method, respectively. We have developed and implemented novel methodologies in the *BoSSS* framework, in which we have refactored and extended the *Compressible Navier-Stokes (CNS)* solver and the *XDGShock* solver in particular. In this section, we summarize the main aspects and novelties of both approaches.

In Chapter 1, this work has been put into a wider context by discussing the open tasks which need to be addressed. In Chapter 2, we have introduced the relevant mathematical and physical theory of supersonic compressible flow, which inherently contains discontinuous flow phenomena such as shock waves. These phenomena are difficult to model and numerically challenging. In Chapter 3, the spatial discretization of the employed unfitted DG methods has been presented (Kummer, 2016; Müller et al., 2017). Among others, we have reported the essential parts of a cell-agglomeration technique, which optimizes the conditioning of the system matrix and lowers the time-step restriction in the context of explicit time-integration schemes for small and ill-shaped cut-cells.

Shock-capturing on cut-cells using a discontinuous Galerkin immersed boundary method[19]
The first of the two numerical approaches developed during the course of this work makes use of a DG IBM, where the considered geometry is represented sharply by the zero iso-contour of

---

[19]Extended version of Geisenhofer et al. (2019, Section 7).

a level-set function. In Chapter 5, we have extended the DG IBM as proposed by Müller et al. (2017) by means of a two-step shock-capturing strategy based on a modal-decay detection of trouble cells and a subsequent smoothing by adding C^0-continuous artificial viscosity (Persson, 2013). In Chapter 4, we have derived an adaptive local time-stepping (LTS) scheme (Winters and Kopriva, 2014; Krämer-Eis, 2017) for the usage on an agglomerated cut-cell grid and combined it with the DG IBM and the presented shock-capturing strategy.

Non-agglomerated cut-cells with activated artificial viscosity limit the maximum admissible time-step size to prohibitively small values in the context of explicit time-integration schemes. This is due to the fact that the time-step size in non-agglomerated cut-cells can differ up to two orders of magnitude compared to standard cells on the background grid in the test cases of this work. The difference may be even larger on locally refined grids. By contrast, we have only considered equidistant grids in this work. Without applying further numerical techniques, the global time-step size would be restricted drastically. As a remedy, our adaptive LTS scheme groups the cells into clusters according to their local time-step sizes and updates each cluster individually in time. An essential feature is the time-dependent reconstruction of the cell clustering so that the physical changes in unsteady flows are considered. This is a novelty compared to the standard application scenario, where LTS schemes are usually applied in the context of adaptive mesh refinement techniques, where the cell clustering is based on a characteristic grid size, see Krivodonova (2010) and references therein.

In Chapter 5, the effectiveness of our combined approach has been verified in several numerical benchmarks in terms of robustness, stability, and accuracy. For example, we have compared our numerical results with the exact solution of the Sod shock tube problem (Sod, 1978) for several polynomial degrees $P \geq 2$ in a boundary-fitted and an immersed-boundary configuration. In both configurations, the same error levels have been obtained as for a global time-stepping scheme while saving up to $63 - 70\%$ of the total cell updates depending on the adaptive LTS reclustering interval. Additionally, we have investigated complex two-dimensional test cases, such as a shock-vortex interaction and a double Mach reflection (DMR), for which similar savings could be obtained. The DMR test case has verified the geometrical flexibility of the presented cut-cell DG IBM.

In summary, the adaptive LTS scheme in combination with the cell-agglomeration technique significantly decreases the otherwise large computational costs while not introducing additional spatial or temporal errors. To the best of our knowledge, we were the first to combine such an approach with a shock-capturing strategy based on artificial viscosity successfully in the context of a DG IBM (Geisenhofer et al., 2019). Thus, this can be considered one of the main contributions of this work.

High-order shock-fitting using an extended discontinuous Galerkin method   The second numerical approach of this work tackles the issue of regaining the global high-order convergence, which is generally lost when using a shock-capturing strategy based on artificial viscosity. We aim to develop an XDG method for supersonic compressible flow, where, besides the solid body, the shock front is represented by a sharp interface description. This work marks the starting point for this endeavor by building a fundamental basis, which we have presented in Chapter 6. To the best of our knowledge, no efforts have yet been made towards this direction in the literature. An essential part is the reconstruction of a level-set function, whose zero iso-contour represents the shock front. For that, we have developed a reconstruction procedure which consists of four steps, see Section 6.3:

(i) We compute the underlying test case by means of a DG method or DG IBM with a shock-capturing strategy based on artificial viscosity as presented in Chapter 5.

(ii) The inflection points of the smoothed density field are considered as a sufficiently exact approximation of the shock front. For that, we extract a set of candidate points by means of a point-based search.

(iii) The candidate points are clustered and iteratively sorted out until we obtain the sought-after inflection points.

(iv) From the inflection points, we reconstruct a C^0-continuous level-set function, whose zero iso-contour approximates the shape and the position of the shock front.

During the reconstruction procedure, we have applied patch-recovery filters (Kummer and Warburton, 2016) and a continuity projection (Smuda, 2020; Smuda and Kummer, 2020) in order to improve the quality of the derived DG approximations and the shock level-set function. It has turned out that this generic methodology is only capable of delivering a *cell-accurate* reconstruction of the shock front. Thus, further efforts have been made in the development of a correction procedure, which is capable of determining the *sub-cell accurate* shock position, see Section 6.4. In this process, we have made use of an implicit pseudo-time-stepping procedure, where we have solved the Euler equations by an implicit Euler scheme with finite time-step sizes, which can be seen as an under-relaxation-like scheme. We have shown a pseudo-two-dimensional proof of concept for a stationary shock wave in Section 6.4. This completely novel methodology can be considered another main contribution of this work.

In addition to the reconstruction procedure of the shock front, we have verified the implementation and application of two level-set functions, which describe a solid body and a shock front, respectively, in Section 6.3.1. Additionally, small and ill-shaped cut-cells of both interfaces are treated by the cell-agglomeration technique presented in Section 3.3.2. The developed code base is capable of handling two level-set functions, which is essential for future general applications in two and three dimensions.

Main contributions   In the following, we briefly summarize the novelties of this work:

- We have developed a DG IBM with a two-step shock-capturing strategy consisting of a modal-decay detection and a smoothing based on artificial viscosity for the application on an agglomerated cut-cell grid. The computational efficiency has been improved by an adaptive LTS scheme with a dynamic reclustering in time. This novel combined approach has been published in Geisenhofer et al. (2019).

- We have built the fundamental basis of an XDG method for supersonic compressible flow, where the zero iso-contours of two different level-set functions are employed for a sharp description of solid bodies and discontinuous flow phenomena. In particular, we have developed a novel sub-cell accurate reconstruction procedure of the shock front by means of an implicit pseudo-time-stepping procedure.

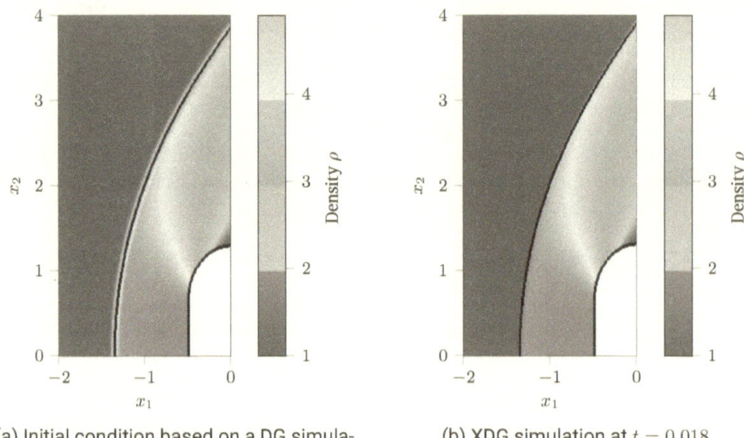

<table>
<tr><td>(a) Initial condition based on a DG simulation with shock-capturing.</td><td>(b) XDG simulation at $t = 0.018$.</td></tr>
</table>

Figure 7.1: Supersonic blunt-body problem computed with an XDG method. The blunt body and the reconstructed shock front are represented by the zero iso-contours of different level-set functions. The XDG simulation with a polynomial degree of $P = 2$ fails after several time-steps, since instabilities arise at the shock front. As a remedy, the sub-cell accurate correction procedure, see Section 6.4, may be adapted to higher dimensions in combination with a preconditioned implicit time-integration scheme. Only the upper half of the domain is plotted for better visibility.

7.2 Outlook

In this section, we briefly discuss possible future work related to *shock-capturing* and *shock-fitting* approaches in the context of DG methods.

Extension of the sub-cell accurate reconstruction of the shock front to higher dimensions
As a final result of this work, we present an XDG computation of a supersonic blunt-body problem with a polynomial degree of $P = 2$ on a grid consisting of 80×320 cells. The shock level-set function is reconstructed based on a converged DG IBM simulation with shock-capturing. The reader is referred to Section 6.3 for details about the applied methodology and a description of the test case. We use the explicit Euler scheme to advance the solution in time. In Figure 7.1, we show the density distribution at the beginning of the simulation and after several time-steps at $t = 0.018$. Even without performing a quantitative assessment, Figure 7.1 clearly indicates that the shock sharpens during the simulation run. However, instabilities arise at the shock and lead to a failure of the simulation for $t > 0.018$. This shows the need of further stabilization and correction mechanisms in order to obtain a robust high-order XDG method for the general higher-dimensional case. One potential remedy may be the sub-cell accurate correction procedure of the shock position, for which we have shown a one-dimensional proof of concept in Section 6.4.3. In two dimensions, each marker point of the level-set function could

be moved along the density gradient by means of the implicit pseudo-time-stepping procedure until it converges to the exact position on the shock. As soon as all marker points are converged, a new shock level-set function could be reconstructed and the computation of the flow field could be repeated. Furthermore, implicit time-integration schemes may be applied to compute the flow field of this steady-state problem. Suitable preconditioning techniques have to be considered in order to solve the nonlinear algebraic equation system in the higher-dimensional case. At this point in time, the *XDGShock* solver is capable of solving small two-dimensional problems by applying an implicit Euler scheme, which uses Newton's method.

Shock-tracking in unsteady flows In Chapter 6, we have only considered steady flows in the context of shock-fitting, where the solid body and the shock front are fixed in space. As soon as unsteady flow scenarios are of interest, it is not foreseeable whether additional techniques are required in order to stabilize the moving shock front. The problem of extending a quantity from a sub-manifold embedded into the computational domain also starts to play a role for the accuracy and the stability of the numerical method (Utz and Kummer, 2017; Utz, 2018).

The discontinuous Galerkin immersed boundary method with shock-capturing on the way to three dimensions In general, the geometrical flexibility of an immersed boundary method (IBM) is favorable for the application in three dimensions, since the effort of grid generation can be reduced significantly. However, the modified Hierarchical Moment-Fitting (HMF) cut-cell quadrature (Müller et al., 2013; Müller et al., 2017) becomes a challenging issue, since the bounding box for the seeding of the quadrature points has to be adapted (Krämer-Eis, 2017), see also Section 3.3.3. An alternative are integration rules based on the work by Saye (2015), tailored to the *BoSSS* framework by Beck (2018). Another challenging issue is the dynamic (de-)agglomeration of cut-cells in the context of two-dimensional sub-manifolds on three-dimensional computational grids, since the complexity of neighboring relations between cut-cells increases (Kummer et al., 2018). The extension of the two-step shock-capturing strategy to three dimensions is a further issue which needs to be addressed. In general, there is no theoretical limitation of constructing a C^0-continuous artificial viscosity field in higher dimensions. However, the implementation may be elaborate due to more complex geometrical relations, in particular at solid bodies where artificial viscosity is usually added.

Dynamic load balancing in the context of local time-stepping schemes The adaptive LTS scheme, presented in Chapter 4, may be implemented for three-dimensional applications. Krämer-Eis (2017) already suggested this extension in the context of a DG IBM, since the computational speed-up may be even larger in the three dimensions. This is due to the fact that the number of cut-cells, which impose a strong restriction on the time-step size in combination with artificial viscosity, grows quadratically, while the total number of cells grows cubically. As a result, a suitable dynamic load balancing technique becomes even more an issue to tackle. The investigations in Section 4.4 have revealed that there is no simple solution to an efficient grid-partitioning when applying the adaptive LTS scheme in the context of unsteady flows (Weber, 2018). A suitable cell cost estimation needs to be developed in order to make this approach compatible. Otherwise, an efficient parallelization will be limited to simple, manually tuned application scenarios.

Long-term goal   In the future, the presented XDG method may become a method of choice if the novel sub-cell accurate reconstruction procedure of the shock front is extended to the general higher-dimensional case. Furthermore, implicit time-integration schemes seem to be fundamentally essential when aiming for non-oscillating numerical solutions with high accuracy in this research area. Additional stabilization mechanisms may be required as soon as unsteady test cases with moving shock waves become a subject of interest.

In summary, the contributions of this work are significant steps towards a high-order XDG method for the simulation of supersonic compressible flow in three dimensions, where discontinuous flow phenomena are represented by a sharp interface description.

A Appendix

A.1 Quadrature Scheme *Gauss and Stokes Preserving One-Step HMF*

In this work, we also use the modified Hierarchical Moment-Fitting (HMF) quadrature variant *Gauss and Stokes Preserving One-Step HMF* as proposed by Kummer (2016) and Smuda (2020). In an extended discontinuous Galerkin (XDG) method, the integrals (3.82) may be evaluated for the basis $\phi_j(\mathbf{x}_l) = \phi_{j,n}(\mathbf{x}_l)$ on a reference cell $K_j \in \mathbb{R}^2$. By applying the HMF scheme, we determine two sets of weights $\tilde{\mathcal{W}}^{\mathfrak{A}}$ and $\tilde{\mathcal{W}}^{\mathfrak{B}}$ for the quadrature over the species volumes and additionally, the set of weights $\tilde{\mathcal{W}}^{\mathfrak{J}}$ for the quadrature over the interface. In a discontinuous Galerkin immersed boundary method (DG IBM), the volume quadrature is further restricted to the fluid region $\mathfrak{A}$. We assume that a suitable set of quadrature nodes $\mathcal{X}$ is given. As for the original variant, we calculate the weights $\tilde{\mathcal{W}}^{\mathfrak{A}}$ and $\tilde{\mathcal{W}}^{\mathfrak{J}}$ by finding the least-squares solution of the under-determined linear system (3.83) build on the Gauss theorem

$$\int_{(\mathcal{X},\tilde{\mathcal{W}}^{\mathfrak{A}})}^{\mathrm{num}} \nabla \cdot \phi_j - \oint_{(\mathcal{X},\tilde{\mathcal{W}}^{\mathfrak{J}})}^{\mathrm{num}} \phi_j \cdot \mathbf{n}_{\mathfrak{J}} = \oint_{\partial K_{\mathfrak{A}}} \phi_j \cdot \mathbf{n}_{\partial K_j} \, \mathrm{d}S \tag{A.1}$$

and on the Stokes theorem

$$\oint_{(\mathcal{X},\tilde{\mathcal{W}}^{\mathfrak{J}})}^{\mathrm{num}} \tilde{\kappa} \mathbf{n}_{\mathfrak{J}} \cdot \phi_j - (\mathbf{I} - \mathbf{n}_{\mathfrak{J}} \otimes \mathbf{n}_{\mathfrak{J}}) : \nabla \phi_j = - \int_{\partial(\mathfrak{J} \cap K_j)} \mathbf{s} \cdot \phi_j \, \mathrm{d}l \,. \tag{A.2}$$

In Equation (A.2), $\tilde{\kappa} = \nabla \cdot (\nabla \varphi / |\nabla \varphi|)$ denotes the curvature of the level-set function $\varphi(\mathbf{x})$ and $\mathbf{s}$ is the tangent of the interface $\mathfrak{J}$.

For a better understanding, we neglect the volume integral $\int_{(\mathcal{X},\tilde{\mathcal{W}}^{\mathfrak{A}})}^{\mathrm{num}} \nabla \cdot \phi_j$ in Equation (A.1) for a moment and construct a surface rule. Note that Equation (A.1) is still a linear system, where we want to determine the set of weights $\tilde{\mathcal{W}}^{\mathfrak{J}}$ with the corresponding vector $\mathbf{n}_{\mathfrak{J}}$. For a specific vector of basis functions $\phi_j \in \mathbb{P}_P(\{K_j\})^2$ with $\nabla \cdot \phi_j = 0$ the corresponding relation to ϕ_j is given by

$$[\phi_j(\mathbf{x}_1) \cdot \mathbf{n}_{\mathfrak{J}}, \ldots, \phi_j(\mathbf{x}_L) \cdot \mathbf{n}_{\mathfrak{J}}] \cdot \tilde{w}^{\mathfrak{J}} = \oint_{\partial K_{\mathfrak{A}}} \phi_j \cdot \mathbf{n}_{\partial K_j} \, \mathrm{d}S = - \oint_{(\mathcal{X},\tilde{\mathcal{W}}^{\mathfrak{J}})}^{\mathrm{num}} \phi_j \cdot \mathbf{n}_{\mathfrak{J}} \,. \tag{A.3}$$

The reader is referred to Equations (3.83) to (3.85) for an introduction of the variables in the context of the moment-fitting system. In particular, $\tilde{w}^{\mathfrak{J}}$ denotes the vector of the unknown weights in Equation (A.3). This relation can be assembled for each basis function of the test space so that we finally obtain the linear system (3.85) as introduced in Equation (3.85).

Finally, the quadrature weights for the volume integration in species $\mathcal{B}$ can be constructed via a complementary approach

$$\int_{K_{j,\mathcal{B}}} \phi_{j,n}\, \mathrm{d}V = \int_{K_j} \phi_{j,n}\, \mathrm{d}V - \int_{K_{j,\mathcal{A}}} \phi_{j,n}\, \mathrm{d}V\,, \tag{A.4}$$

or, alternatively, can be calculated independently by using Equations (A.1) and (A.2).

A.2 Simulation Parameters

In Tables A.1 and A.2, we list the simulation parameters of the test cases presented in Chapters 5 and 6, respectively, for the sake of reproducibility.

#	Test case	Sec.	Fig./Tab.	Ω	h	P	t_{end}/time-steps
1	Scalar transport equation	5.4.1	Fig. 5.2	$[0,1] \times [0,1]$	0.1	2	$\{0, 16, 3500\}$
2	Scalar transport equation	5.4.1	Fig. 5.3	$[0,1] \times [0,1]$	0.1	2	$\{0, 0.05, 0.25\}$
3	Scalar transport equation	5.4.1	Fig. 5.4	$[0,1] \times [0,1]$	$\{0.1, 0.05, 0.025\}$	$\{2, 4, 10\}$	0.25
4	Inviscid Burgers' equation	5.4.1	Fig. 5.5	$[0,1] \times [0,1]$	0.1	10	$\{0, 0.25, 0.5\}$
5	Scalar transport equation	5.4.2	Fig. 5.6	$[0,1] \times [0,1]$	$\{0.1, 0.05\}$	$\{2, 3, 4, 5, 6, 7, 8\}$	0.5
6	Sod shock tube	5.5.1	Fig. 5.7	$[0,1] \times [0,1]$	0.02	$\{2, 3, 4\}$	0.25
7	Sod shock tube	5.5.1	Fig. 5.8	$[0,1] \times [0,1]$	0.02	$\{2, 3, 4\}$	$\{\approx 0.09, \approx 0.21, 0.25\}$
8	Sod shock tube	5.5.1	Fig. 5.9	$[0,1] \times [0,1]$	0.02	$\{2, 3, 4\}$	0.25
9	Sod shock tube	5.5.1	Tab. 5.1	$[0,1] \times [0,1]$	0.02	$\{2, 3, 4\}$	25
10	Shock-vortex interaction	5.5.2	Fig. 5.10	$[0,2] \times [0,1]$	$0.00\overline{3}$	3	$\{0.16, 0.7\}$
11	Shock-vortex interaction	5.5.2	Fig. 5.11	$[0,2] \times [0,1]$	$0.00\overline{3}$	3	$\{0.16, 0.7\}$
12	IBM Sod shock tube	5.5.3	Fig. 5.12	$[0,1.5] \times [0,1.1]$	0.02	2	0.25
13	IBM Sod shock tube	5.5.3	Fig. 5.13	$[0,1.5] \times [0,1.1]$	$\{0.02, 0.01, 0.005\}$	$\{2, 3, 4\}$	0.25
14	IBM Sod shock tube	5.5.3	Fig. 5.14	$[0,1.5] \times [0,1.1]$	0.02	3	0.25
15	IBM Sod shock tube	5.5.3	Tab. 5.2	$[0,1.5] \times [0,1.1]$	0.02	$\{2, 3, 4\}$	0.25
16	IBM double Mach reflection	5.5.4	Fig. 5.15	$[0,3] \times [0,2]$	0.01	3	0.2
17	IBM double Mach reflection	5.5.4	Fig. 5.16	$[0,3] \times [0,2]$	0.01	3	0.2
18	Double Mach reflection	5.5.4	Fig. 5.16	$[0,4] \times [0,1]$	$0.01\overline{3}$	3	0.2

#	S_0	ε_0	κ	Time-step.	Q	I_{LTS}	C_{init}	δ_{agg}
1	$6.25 \cdot 10^{-4}$	10.0	1.0	RK	1	-	-	-
2	$6.25 \cdot 10^{-4}$	10.0	1.0	RK	1	-	-	-
3	$6.25 \cdot 10^{-4}$	10.0	1.0	RK	1	-	-	-
4	$6.25 \cdot 10^{-4}$	1.0	1.0	RK	1	-	-	-
5	$6.25 \cdot 10^{-4}$	10.0	1.0	RK	1	-	-	-
6	$1.0 \cdot 10^{-3}$	1.0	0.5	AB, LTS	3	1	3	-
7	$1.0 \cdot 10^{-3}$	1.0	0.5	AB, LTS	3	1	3	-
8	$1.0 \cdot 10^{-3}$	1.0	0.5	AB, LTS	3	$\{0, 1, 5, 10, 20, 50\}$	3	-
9	$1.0 \cdot 10^{-3}$	1.0	0.5	AB, LTS	1	1	1	-
10	$1.0 \cdot 10^{-3}$	1.0	0.5	AB, LTS	3	1	3	-
11	$1.0 \cdot 10^{-3}$	1.0	0.5	AB, LTS	3	1	3	-
12	$1.0 \cdot 10^{-3}$	1.0	0.5	AB, LTS	3	1	2	0.3
13	$1.0 \cdot 10^{-3}$	1.0	0.5	AB, LTS	3	1	2	0.3
14	$1.0 \cdot 10^{-3}$	1.0	0.5	AB, LTS	3	$\{0, 1, 5, 10, 20, 50\}$	2	0.3
15	$1.0 \cdot 10^{-3}$	1.0	0.5	AB, LTS	3	1	2	0.3
16	$1.0 \cdot 10^{-4}$	1.0	0.5	AB, LTS	3	1	2	0.3
17	$1.0 \cdot 10^{-4}$	1.0	0.5	AB	3	-	-	0.3
18	$1.0 \cdot 10^{-4}$	1.0	0.5	AB	3	-	-	-

Table A.1: Simulation parameters of the test cases presented in Chapter 5. We scale the shock-capturing parameters $S_0 \sim 1/P^4$ and $\varepsilon_0 \sim h/P$ if not explicitly stated.

#	Test case	Sec.	Fig./Tab.	Ω_i	h	P	$t_{\text{end}}/\Delta t$
1	Supersonic blunt body problem	6.3, 6.3.4	Figs. 6.4, 6.8, 6.9, 6.10, 6.15	$[-2, 0] \times [-4, 4]$	0.05	2	16.0
2	Gaussian bump	6.3.1	Figs. 6.5, 6.6	$[-20, 20] \times [0, 20]$	$\{0.3125, 0.625, 1.25\}$	$\{0, 1, 2, 3\}$	-
3	Stationary normal shock wave	6.4.2	Fig. 6.11	$[0, 1] \times [0, 0.3]$	0.1	2	$\Delta t = 0.1$
4	Stationary normal shock wave	6.4.3	Fig. 6.12	$[0, 1] \times [0, 0.3]$	0.1	$\{2, 4\}$	2.0
5	Stationary normal shock wave	6.4.3	Fig. 6.13	$[0, 1] \times [0, 0.3]$	0.1	2	-
6	Stationary normal shock wave	6.4.3	Fig. 6.14	$[0, 1] \times [0, 0.3]$	0.1	2	$\Delta t = 0.1$

#	S_0	ε_0	κ	Time-step.	Q	δ_{agg}
1	$1.0 \cdot 10^{-3}$	1.0	1.0	RK	1	0.3
2	-	-	-	RK	1	0.3
3	-	-	-	BDF	1	0.3
4	$1.0 \cdot 10^{-3}$	1.0	0.5	RK	1	0.3
5	-	-	-	-	-	0.3
6	-	-	-	BDF	1	0.3

Table A.2: Simulation parameters of the test cases presented in Chapter 6.